The Politics of Maps

The Politics of Maps

Cartographic Constructions of Israel/Palestine

CHRISTINE LEUENBERGER
AND IZHAK SCHNELL

OXFORD
UNIVERSITY PRESS

OXFORD
UNIVERSITY PRESS

Oxford University Press is a department of the University of Oxford. It furthers the University's objective of excellence in research, scholarship, and education by publishing worldwide. Oxford is a registered trade mark of Oxford University Press in the UK and certain other countries.

Published in the United States of America by Oxford University Press
198 Madison Avenue, New York, NY 10016, United States of America.

Library of Congress Cataloging-in-Publication Data
Names: Leuenberger, Christine, author. | Schnell, Izhak author.
Title: The politics of maps : cartographic constructions of Israel/Palestine /
Christine Leuenberger, Cornell University,
and Izhak Schnell, Tel Aviv University.
Description: New York, NY : Oxford University Press, [2020] |
Includes bibliographical references and index.
Identifiers: LCCN 2019057040 (print) | LCCN 2019057041 (ebook) |
ISBN 9780190076238 (hardback) | ISBN 9780190076252 (epub)
Subjects: LCSH: Cartography—Political aspects—Israel. |
Cartography—Politcal aspects—Palestine.
Classification: LCC GA1323.7.A1 L48 2020 (print) |
LCC GA1323.7.A1 (ebook) | DDC 956.9405—dc23
LC record available at https://lccn.loc.gov/2019057040
LC ebook record available at https://lccn.loc.gov/2019057041

3 5 7 9 8 6 4 2

Printed by Integrated Books International, United States of America

To my beloved children Annika Caroline Pinch and Benika Joan Pinch
Christine

To my beloved family Gila, Yael, Michal, and Zohar
Izhak

Contents

Foreword

How is it that an Israeli geographer and a Swiss-American sociologist and Science Studies scholar came together to write about the politics of maps in Israel? It all started with an email in 2005. At that time, I (Christine) asked then graduate student Shay David whether he could assist me in finding Israeli academics interested in co-investigating the social consequences of the "Separation Wall" that the Israeli government had started to build in 2002. He connected me with Uri Gopher, who wrote to me on October 14, "I am more than happy to assist you with your interesting and important research initiative" and proceeded to name Izhak from the Department of Geography and the Human Environment at Tel Aviv University as a professor who "is very keen to speak with you about this." By November 2, Izhak and I were already brainstorming about what that joint project could be. Because of my ongoing teaching obligations and lack of funding, it was not until spring 2008, after I received a Fulbright Scholar Award, that I could spend a semester at Tel Aviv University.

Upon my arrival in Tel Aviv, Izhak became one of my most trusted informants, friends, and colleagues. As we traveled together across Israel and the West Bank, we kept talking about the history and geography of the Israeli-Palestinian conflict. It became increasingly clear to me that in order to fathom the complexity of the geography of the conflict, one had to walk the land and drive along and through it to fully appreciate the intermingling of Palestinian and Jewish Israeli spaces. Indeed, the embodied experience of being in these spaces of contention can provide a most vivid and vicarious understanding of this entrenched conflict over land, territory, and histories.

Ethnographers have long maintained that having a trusted local mentor is the single most important aspect of doing fieldwork. Izhak opened my eyes to the complexities of Israeli society in the midst of the continuously simmering Israeli-Palestinian conflict. The land- and cityscapes became alive with the stories he told. The intricate social landscapes of Israeli society, its divisions and fissures, but also its unity and coherence became ever clearer to me as we were sipping coffees, walking along the barrier's route, and driving for hours through the plains and hills of Israel/Palestine. During those many hours, Izhak would share his knowledge, stories, and experiences from having lived in a divided land all his life, and I was always eager to listen, to learn, and to understand what for me was uncharted territory. It was on one of our many walks to lunch in one of Tel Aviv University's cafeterias that we started talking about maps.

There were many maps hanging on the walls of Tel Aviv's geography department. They all differed from one another, and I had become intrigued by their differences, by their textures, and by what they cartographically represented and excluded. Yet it was not until one day in May 2008 that I started to appreciate the power of the politics of maps. That day I was to meet Abir,[1] a Palestinian journalist, who was to drive me into the West Bank to interview several Palestinian experts. She was to meet me outside the American Colony Hotel in East Jerusalem. Armed with an Israeli-produced Carta map of the West Bank, I was ready to trace our route through what for me was then still unknown territory. Abir skillfully navigated through a maze of streets, avoiding the many checkpoints into the West Bank, and drove through an unfinished section of the Separation Wall crossing from a Jewish to a Palestinian neighborhood. There we were—in Palestine. Then, suddenly, Abir turned to me and explained that there had been a shooting at the notorious Kalandia checkpoint between Jerusalem and Ramallah with people injured. My heart fell like a stone. Here I was in the West Bank and things were heating up at the checkpoints. But Abir said, "Don't worry; we will stay out of the claws of the lion." I wasn't quite sure what she meant but trusted that she knew how to avoid the hot spots. Yet my trust in her did not stop me from following our route on my Carta map that I had purchased in Tel Aviv.

On this map, some of the roads were marked green (which in the map's key were described as safe roads for Israeli drivers); others were marked red (apparently to be avoided by Israeli drivers). Yet, as we were driving down a winding, badly maintained, road I suddenly could no longer match our location to the map at all. Slightly panicked, not least because we were to stay out of the "tiger's claws," I tried—for what seemed like a rather long time—to trace our location. Finally, I plugged up the courage to tell Abir "I cannot find where we are on the map anymore." She looked at me and laughed. "What is this map you have there? Oh, a Carta map. No wonder—we came off that map a long time ago!" Not only was I now in unknown territory, but we had also just fallen off the map. I wondered, how could we just have fallen off the map? Why were Israeli infrastructures in the West Bank over-represented while Palestinian roads, towns, and villages were under-represented or eliminated from the map? The reasons for our falling off the map that day was to preoccupy me for the next few years.

It was on one of my and Izhak's many walks to one of Tel Aviv University's cafeterias that Izhak told me the story of how he also had become fascinated by the politics of maps. As he recounted, it was 2005 and an Israeli withdrawal from the Gaza Strip was imminent. So, he will pick up the story from here.

[1] My guide's name has been changed.

With public debate over the move raging, I (Izhak)—as a member of the Israeli non-governmental organization (NGO) Peace Now[2] (which advocates a two-state solution to the Israeli-Palestinian conflict)—was tasked to produce a set of maps. After the then prime minister Ariel Sharon had suggested a possible Israeli withdrawal from Gaza and Jericho, Peace Now conducted a survey to learn Israeli settlers' possible response to a withdrawal. The data showed that to do so would not cause a widely feared civil war in Israel. The survey was of interest to the media and I started to prepare two geographic information systems (GIS) maps visualizing the data. The maps showed three different possible responses by the settlers to a withdrawal: acceptance of compensation for withdrawal, opposition, or outright violence.

While working on these maps I discovered that someone had broken into my office and stolen some of my computer disks. I did not connect this incident to the fact that I had been commissioned to produce these contentious and politically sensitive maps until a few days later, when my computer used to design the maps, was also stolen!

At the time, journalists and settler advocates contacted me, demanding access to the information ahead of publication. The survey and the maps were eventually published in one of Israel's most influential newspaper, *Haaretz*. A few weeks later I was invited to give a lecture to veteran military officers, who referred me to the head of the Council for National Security, which was planning the withdrawal from Gaza and Jericho. After my presentation to the Council, I was told that the withdrawal could not be planned based on the maps' apparently overly optimistic assumptions. Yet, the withdrawal took place nevertheless. The maps had become part of the public debate and the political decision making at the time. However, the map thieves were never caught. Strangely, I felt weirdly pleased that after years of conducting academic research, my work had suddenly seemed important enough to be stolen. It was then that I started to appreciate the political power of maps.

Years later, in 2010, I was asked by the Steinmetz Center for Peace Studies at Tel Aviv University to propose a conference that connects geography to peace research. At that time, Christine and I had just started to publish our first results on the politics of maps in Israel (Leuenberger and Schnell 2010). After having long been neglected by geographers, political aspects of maps had started to gain attention. This conference provided a forum for Israeli scholars and cartographers to systematically and critically question the politics of maps in Israel for the first time (Schnell and Lavie 2012). So, this is how our (Christine's and Izhak's) shared story began.

[2] See more information about Peace Now at http://peacenow.org.il/eng/content/who-we-are.

While it is a common assumption that scientific collaborations and projects are well planned and motivated by the most pressing problems of the day, our long-term collaboration exemplifies what other researchers have often found: the importance of circumstantial factors, lack of planning, and the power of serendipity. Indeed, this project started off with an unexpected email in 2005. At that time, we had planned to work on walls and barriers, but after we both had become fascinated and deeply troubled by maps in different ways, we ended up working on maps. We started to collect maps, poring over them and trying to decipher differences of representational strategies between them. We had found our shared passion and interest! It was not walls, but it was space, and the politics of maps.

Yet we knew that whether writing about walls or maps in Israel/Palestine, doing so would be like walking through a political minefield. Indeed, it is astonishing that—when it comes to the Israeli-Palestinian conflict—most everyone seems to feel entitled to have very strong opinions on the matter. Protagonists readily pass judgments on either side and show their partisan support based on often little or misleading information. Moreover, the politics surrounding the conflict has become so taken-for-granted, polarized, and hardened that it is virtually impossible to approach issues apart from their political fossilization. This book aims not to "grind an evaluative ax" (P. Scott, Richards, and Martin 1990) but to provide what anthropologist Clifford Geertz (1973) called a "thick description" by uncovering social processes shaping knowledge claims, and describing and understanding actions, motives, and histories in their respective social, political, and cultural context.

Acknowledgments

A project that has been in the making for many years is more than the product of its authors. We are indebted to so many communities of practitioners, scholars, friends, and family that they are too numerous to mention all. We are most immediately indebted to colleagues and friends, as well as institutional support from our respective departments, Tel Aviv's Geography Department and Cornell University's Department of Science and Technology Studies. Christine is also indebted to her hosts: Jad Isaac at the Applied Research Institute Jerusalem (ARIJ); Stefan Ziegler at the United Nations Relief Works Agency (UNRWA); Eitan Alimi and Ruth Kark at Hebrew University. She also remembers fondly the friends and colleagues at Tel Aviv University's geography department as well as Izhak's family Gila, Yael, Michal, and Zohar who all became extended family. There are always some key people who provide support, friendship, and help throughout and who make such a long-term engagement worthwhile; these include, of course, her co-author Izhak Schnell, Snait Gissis, Ahmad El-Atrash, Oren Yiftachel, and Stefan Ziegler.

For many conversations, inspirations, discussions, and friendships, Christine is also indebted to Hillel Schenker, Moshe Brawer, Efrat Ben Zeev, Ilil Naveh-Benjamin, Anat Leibler, Ariane Littman, Ariel Handel, Gideon Biger, Haim Srebro, Walid Salem, Cedric Parizot, Dani Filc, Dafne Hirsch, Daniel Bar-Tal, Fadwa Azem, Fida Abdel-Latif, Fuad Hudali, Khalil Tufakji, Suheil Khalilieh, Issa Zboun, Ibrahim Jaffal, Bernhard Sabella, Sami Adwan, Jessica Montell, Jose Brunner, Samira Alayan, Kathleen Vogel, Sholpan Butterfield, Philip Butterfield, John Cutliffe, Jess Bier, Alex Koensler, Karen Culcasi, Linda Quiquivix, Marwan Tarazi, Maya Rosenfeld, Nicolas Pelham, Nimrod Bar-Am, Nissim Mizrachi, Noah Efron, Ruchama Marton, Ruthi Sharon, Smadar Sharon, Dafne Hirsch, Sami Adwan, Talia Fried, Yaniv Fisher, Yuval Yonay, Anat Ben-David, Nick Ninos, Asaf Darr, and Eran Fisher. For warning her of in-coming missiles during two wars in 2012 and 2014 and always having her safety at heart, she would like to thank her guardian angels: Izhak Schnell and Miron M. Denan.

For academic and personal engagement, support, and encouragement of different kinds she is grateful to Luke Struckman, Tristan Sturm, Sidney Tarrow, Thaddeus Müller, Jerome Bourdon, Reece Jones, Zahraa Zawawi, Marcello Di Cintio, Elisabeth Vallet, Mitch Albert, Olivia Snaije, Ethan Tidwell, Raz Chen-Morris, Emily Regan Wills, Brigitte Piquard, Salim Tamari, Abdullah Herzallah, Anne-Laure Amilhat Szary, Jawad Fatayer, Philippe Rekacewicz, Samer Alatout,

Oren Livio, Nurit Peled-Elhanan, Gary Fields, Kenneth Madsen, Julien Chippone-Luccesi, Olivier Legrand, Dana Rassas, Hubert Knoblauch, Thomas Luckmann, Simone Tosoni, David Newman, Ghazi-Walid Falah, George Lowery, Richard Kiely, Tally Kritzman-Amir, Ronit Donyets-Kedar, Asher Kaufman, Stephan Hilgartner, Judith Reppy, Jay Jasanoff, Tal Golan, Yaakov Garb, Yossef Schwartz, Josh Mitnick, Nuha Musleh, Shay David, Micha David, and Uri Gopher. For access to map collections Christine is thankful to Tel Aviv University's geography department, the United Nations Office for the Coordination of Humanitarian Affairs (UNOCHA), Birzeit University, the Applied Research Institute Jerusalem (ARIJ), the Palestinian Ministry of Planning, the Palestinian Central Bureau of Statistics (PCBS), Harvard University, the Jabutinsky Institute, the Survey of Israel, and the Arab Studies Society. Christine and Izhak are also grateful to Peter Wissoker and Ranjit Singh for helping to finalize the manuscript.

Furthermore, Christine is grateful for research support from the Fulbright Scholar Program for the Middle East, North Africa, Central and South Asia Regional Research Program (grantee no. G48413539), the Fulbright Specialist Program (grantee no. 88100362), and the National Science Foundation (award no. 1152322). She also is thankful to her family, Verena Leuenberger, Theodor Leuenberger, and Trevor Pinch. Most of all she is indebted to her children Benika Joan Pinch and Annika Caroline Pinch who endured their mother's absence for weeks and months at a time with strength, endurance, and a hard-won appreciation of why this project had become so important to her. Izhak is also grateful to his family: Gila, Yael, Michal, and Zohar. This book is dedicated to both our families.

1
Why Maps Matter

Introduction

Maps are ubiquitous. Climate maps inform us of weather patterns. Political maps designate nations, boundaries, and state capitals. Economic maps reveal a region's natural resources and economic activities. During presidential elections in the United States, maps delineate "blue" Democratic and "red" Republican states. Maps—whether they depict weather patterns, national boundaries, or election results—help us understand and navigate the world around us. Maps don't just come on paper; they also come with iPhone applications. Drivers increasingly rely solely on global positioning systems (GPS) to get to their destinations. Online software, such as Google Earth, depicts spatial configurations and provides satellite images of most localities. Geographic information systems (GIS), such as ArcGIS, ArcView, and Manifold, enable scholars, laypeople, and policymakers to design geographic, thematic, or analytic maps and visualize data cartographically. Moreover, mapping software and phone apps are providing the forum for various stakeholders and interest groups to engage in "app-based activism" (Ackerman 2010, 1), using maps as visual tools to persuade and appeal to wider publics and disseminate geopolitical visions.

Historically, however, maps were the purview of a trained cadre of experts and surveyors that were in the service of the powerful—feudal lords, royalty, and nation-states. Indeed, it was with the rise of nation-states in the 17th century, that maps, which represent calculable spatial relations within a given territory, became tools to control territories, allocate property rights, serve military needs, and consolidate state power. Maps were powerful because "putting the state on the map meant knowing and imagining it as real—and, so, making it a reality" (Biggs in Agnew 2007b: 401). Maps helped create nations, boundaries, spatial imaginations, and national identities (Crampton 2010b; Edney 1997; Harvey 2009; Jacob 2005; Jones 2008; Kosonen 2008; Krishna 1996; Paasi 1999, 2002; Radcliffe 2010; Winichakul 1994; Wood and Fels 1992; Wood, Fels, and Krygier 2010).

In the later part of the 20th century, however, map-making became transformed from an expert's endeavor into a "people's cartography" (Crampton and Krygier 2005, 18; Monmonier 2006; Radcliffe 2009; Schnell and Leuenberger 2014). The collection of spatial data and the business of map-making has become,

The Politics of Maps. Christine Leuenberger and Izhak Schnell, Oxford University Press (2020).

DOI: 10.1093/oso/9780190076238.001.0001

more than ever, "available to anyone with a home computer and an Internet connection" (Crampton and Krygier 2006: 12). Web-based software, ranging from collaborative mapping projects to open-source mapping applications to Google Map Maker, allows users to make their own maps, gather data, and integrate them into different cartographic layers so as to disseminate "alternative mappings" (Perkins 2004, 388) and circulate "counter-mapping and counter-knowledges" (Crampton 2009: 91). Such unprecedented access to map-making technologies provides new forums for political action and for shaping geopolitical visions (Edwards 2009). The internet, in particular, has become a site for various social groups and grassroots movements to disseminate alternative mappings. For instance, the non-governmental organization (NGO) "Americans for Peace Now"[1] has developed an application downloadable to iPhones that consists of Jewish settlements on a scalable Google-built map of the West Bank and Jerusalem. The aim is to disseminate information about the spread of illegal Jewish settlements in the Occupied Palestinian Territories (oPt), and to pressure Israel to halt settlements construction.[2] Also a "My Palestine" app pinpoints Palestinian heritage sites and raises awareness of the Israeli occupation of the Palestinian Territories. App stores are therefore increasingly becoming a new battleground over contested territories.

Such maps frequently challenge officially produced maps by states, providing counter-maps, bottom-up maps, and resistance maps that challenge the traditional wisdom and reveal the silences and omissions that are an inevitable part of maps made in the service of a state. In such an environment, where user-defined maps are increasingly prevalent, it is particularly important to investigate how maps may serve as rhetorical tools to advocate for particular, socially constructed, value-laden, and ideologically charged geopolitical visions (Crampton 2001; Monmonier 2006).

Maps produced in a conflict region, like Israel/Palestine, provide an exemplary case study of alterative mapping practices. This region tells a rich story about the interlinkages between maps and politics. It is a place where territories and boundaries, as well as nation and statehood, remain contested. For instance, since the establishment of the state of Israel in 1948, neighboring countries have

[1] See Americans for Peace Now. Available at http://peacenow.org/. Retrieved Aug 20, 2012.

[2] According to International Law, the Palestinian Territories are under Israeli occupation. The UN Security Council and the International Court of Justice (ICJ) also determined that having Jewish settlements in occupied territories contravenes international law and the Fourth Geneva Convention, which prohibits the transfer of civilian populations into occupied territory. The Israeli government, however, argues that the Fourth Geneva Convention does not apply to the Palestinian Territories and contends that due to Israel's need for self-defense and its replacement of Jordanian rule over the West Bank, its presence in the West Bank has a unique legal character (see, e.g., International Court of Justice. Available at http://www.icj-cij.org/docket/index.php?pr=71&code=mwp&p1=3&p2=4&p3=6, retrieved December 1, 2015). See also Ayoub (2004).

disputed its territorial status and legitimacy. Similarly, Israel's declaration of Jerusalem as its capital is not internationally recognized as, according to international law, East Jerusalem is considered to be under Israeli occupation. While differences about the contours of the national territory had remained latent pre-1967 within Israel, that unanimity evaporated after the 1967 war between Israel and its neighboring states of Egypt, Jordon, and Syria. At that time, Israel more than tripled the territory under its control, as it occupied territories formerly controlled by Jordan (West Bank and East Jerusalem), Egypt (Gaza, the Sinai Peninsula), and Syria (Golan Heights), producing controversies about Israeli territorial claims within as well as outside Israel/Palestine.

The aftermath of the 1967 war also strengthened the political consciousness of Palestinians living in Israel as well as under Israeli occupation in the West Bank and Gaza and in the diaspora; and it gave rise to a neo-Zionist national movement within Israel and the Jewish diaspora that aimed to recover the whole Land of Israel—a biblical Palestine on both sides of the Jordan River (Avineri 1971). The legal territorial quandary post-1967 only intensified the Israeli-Palestinian conflict; the location of boundaries and the territorial status of Israel and the Palestinian Territories became increasingly a matter of public debate, and maps become an integral part of it.

Whether we look at maps produced by the Israeli state or by various governmental or non-governmental organizations or interest groups, they always incorporate and represent wider discourses within Israeli society. Organizations, ranging from (what in Israel are considered) leftist human rights and peace groups to social advocacy groups and political parties associated with the Israeli right, have produced an array of different maps of Israel/Palestine. These maps communicate different narratives, including concerns about Israel's national security and demography; its occupational strategies in the Palestinian Territories; its borders' compliance or non-compliance with international law; and claims pertaining to Jewish rootedness in "Greater Israel" (the area that constituted Palestine prior to the establishment of the Emirate of Jordan in 1922). Palestinian cartographers also use maps with the aim of facilitating the building of a nation-state of Palestine.[3] Various Palestinian mapping genres also reflect map-makers' concern with the Israeli occupation and the concomitant territorial annexation of Palestinian lands, the infringement of international law by the Israeli state, as well as Palestinians' rootedness in the land and their territorial displacement.[4] Various state and non-state Israeli and Palestinian actors therefore use an array

[3] See, e.g., the Statistical Atlas of Palestine. Available at http://www.pcbs.gov.ps/Downloads/book1624/index-e.htm; Leuenberger 2013b.

[4] See, e.g., Applied Research Institute Jerusalem, http://www.arij.org/index.php/geopolitical-maps; Abu-Sitta 2007.

of cartographic strategies to relay these concerns and thereby reflect the politics of maps in societies in intractable conflict.

Understanding Maps: Theories and Methods

Work in critical cartography has long interrogated maps as representations and has pointed to links between maps and politics (Monmonier 1996, 2006). For instance, all maps involve distortion by projecting a three-dimensional curved globe onto a two-dimensional flat surface. Besides selecting a projection, mapmakers also have a choice as to how to construct a map's content. For instance, according to the "rule of ethnocentricity," societies tend to place their own territory at the center of the world map (Harley 1991: 236). Also, the intended audience of a map can co-determine what is included or excluded. During South Africa's apartheid regime, maps tended to foreground white settlements and ignore or minimize black townships, as the maps' target audience was white settlers (Black 1997). Maps are thus designed for certain audiences, and they communicate particular social, cultural, and political assumptions and preferences.

While geographers traditionally differentiated between scientific and political propaganda maps (Agar 1977; Robinson 1952; Tyner 1974), John Pickles (Pickles 1992) argues that to do so has become obsolete. Indeed, not only do *all* maps communicate social, cultural, and political assumptions and preferences, but they also can become tools to shape, legitimize, and institutionalize certain forms of knowledge and collective spatial imaginations. Historically, national movements and states have always used various types of maps to consolidate national images that facilitated the hegemony of the nation-state and helped shape collective spatial imaginations (Gramsci 1971; Hobsbawn 1992; Paasi 1999). For instance, national maps, irrespective of whether they stress or omit cartographic details, tend to emphasize borders, turning the territorial shape into a recognizable logo, which can help create the imagined community of a nation (Anderson 1983; Edwards 2009). Maps can also depict patterns of control and preferred national boundaries, emphasize potential threats and political disputes, and stimulate sociopolitical action.

Such political and social concerns shape the maps' hierarchization of spaces. They impact the visual and linguistic information that is included or omitted, such as the size, color, and symbols as well as the names and narratives incorporated (Harley 1991; Wood and Fels 1992). In the process of mapping, landscapes are thereby reconstituted as socially meaningful signs, subject to selection, classification, abstraction, scale, and simplification (Pickles 2004). Therefore, the "truth" of a landscape is not revealed once social and political factors "distorting" cartographic representations are eliminated. Rather, maps do not reveal, but produce reality; they do not represent, but become inscribed in the landscape; they are not fixed, but are continuously made and re-made. This entails a "shift from ontology (how things are) to ontogenesis (how things become)—from (secure) representation to (unfolding) practice" (Kitchin and Dodge 2007: 335). The focus on the practice of mapping reveals that *all* maps "are necessarily selective, contingent and contextual . . . to solve relational, spatial problems" (Kitchin and Dodge 2007: 343). This shift in critical cartography from understanding maps not as secure representations but as constituted by textual and cartographic practices that are inevitably selective, informed by particular social and historical contexts, and shaped by their function and audience provides us with conceptual tools for interpreting maps.

Critical cartography's focus on the context and practice of mapping also speaks to Science Studies' commitment to studying science in context. Science Studies scholars focus on the political, social, cultural, and institutional factors that impact the sciences. One of the hallmarks of Science Studies has been the study of scientific controversy—the real-time process in which knowledge is constituted while being as yet unfinished. Focusing on controversy has become an analytical way to contextualize knowledge claims and gain unique insight into the production of knowledge and the politics of science (Collins 1981; Collins and Pinch 1993; Jasanoff 2004; Jasanoff et al. 1995; Knorr Cetina 1995; Nelkin 1995; Pinch 1985). Arguably, the geography and cartography of Israel/Palestine is an ideal laboratory for Science Studies. Map-making in this region is a controversy in the making as Israel's national borders continue to be disputed, and Israeli, Palestinian, and internationally produced maps depict the region differently (Biger 2008, 2009; Leuenberger 2013). The history of mapping Israel/Palestine also speaks to how the sciences have become an integral part of nation-state-building in the 20th century, and thereby this study contributes to Science Studies literature on the co-production of science and the state (Carroll 2006; Eghigian, Killen, and Leuenberger 2007; Gerovitch 2001; Jasanoff 2004; Mukerji 2002).

Furthermore, the turn to the "rhetoric of inquiry" in Science Studies (Nelson, Megill, and McCloskey 1987) points to the importance of language and rhetoric in the sciences (Bazerman and Paradis 1991; Billig 1989; McCloskey 1990; Pinch 1996). Indeed as Nelson et al. (1987) note, "Facts themselves are mute. Whatever the facts, *we* do the speaking—whether through them or for them." (p. 8) Scientists, therefore, necessarily use rhetoric to construct the objectivity of their findings, invoke the authority of their expertise, and use analogies and metaphors to communicate effectively. Arguably, cartographers also necessarily use visual forms of rhetoric to do so. How they use various textual and visual rhetorical devices to do so is the subject of our analysis.

While geographers, critical cartographers, and Science Studies scholars have analyzed the visual content of maps (Monmonier 1996; Raj 2007; Rudwick 1976;

Turnbull and Watson 1989; Wood and Fels 2008; Wood and Krygier 2005), methodologically comprehensive and theoretically grounded "rhetorical close reading of maps" are limited (Harley 1991: 242). Only more recently has academic attention started to focus on how visual and textual devices inform images and communicate certain social, cultural, and political messages (Culcasi 2006; Dror 2004; Kosonen 2008). However, despite the importance of empirically analyzing the role, function, and power of visual and textual devices in maps, there still is a lack of a coherent methodological and theoretical framework that can provide an understanding of the content and function of maps. Arguably, linking the conceptual frameworks of Science Studies and Critical Cartography provides a theoretically rich and empirically sensitive framework in order to better understand the politics of maps as it enables us to focus on both the macro domain of politics, culture, and state-making and the micro politics of making maps in practice.

In order to tell the story of the politics of maps in Israel/Palestine we relied on various qualitative methodologies. The data include contemporary and historical cartographic material collected from archival, library, and commercial sources; the internet; the news media; and various governmental and non-governmental organizations and political interest groups. We present various maps from the 1920s to the present that represent a certain generic style of map-making during different historical periods. We also conducted numerous in-depth qualitative interviews with map-makers from governmental and non-governmental organizations, academia, and various political advocacy groups who produce maps for political, academic, or commercial purposes and audiences.

Interviews were conducted with various stakeholders from Israeli governmental and non-governmental institutions, ranging from the Survey of Israel to NGOs, including B'Tselem: The Israeli Information Center for Human Rights in the Occupied Territories; the Geneva Initiative; Zochrot (in Hebrew: Remembering); and the Israeli political party Tafnit and the Jabutinsky Institute. Interviews with Israeli academics were also conducted at Beer Sheva University, Tel Aviv University, Bar Ilan University, and Hebrew University, among others; interviewees from international organizations included leaders of mapping projects funded by the US Agency for International Development (USAID); Middle East experts from the International Crisis Group; and various United Nations (UN)-affiliated cartographers and planners from the United Nations Office for the Coordination of Humanitarian Affairs (UNOCHA) and United Nations Development Programme (UNDP), among others. Palestinian interviewees were affiliated with a range of institutions such as the Palestinian Liberation Organization (PLO) Negotiations Affairs Department, the Palestinian National Authority (PNA)'s Ministry of Transport and Ministry of Tourism, the Palestinian Legislative Council, and various municipalities throughout the Palestinian Territories. Interviews were also held with experts from Jerusalem's Center for Democracy, the GeoMaps Center, the Applied Research Institute Jerusalem (ARIJ), the Palestinian Academic Society for the Study of International Affairs (PASSIA), the Arab Studies Society, and the Palestine Mapping Center; we also talked with cartographers and planners from Bethlehem University, Al-Quds University, and Birzeit University. Interviews provided us with insights into how different map-makers talk about their map-making activities, how they understand them as part of larger sociocultural and political circumstances, and how their map-making practices either converge or diverge from state-enforced territorial claims. As our interviewees would produce a variety of maps, ranging from "official" governmentally sanctioned representations of the territories to "unofficial" maps of the region, they provided us with insights into map-making practices "from above" and "from below" as well as into the institutional, financial, technical, and infrastructural resources that sustain them.[5]

We also conducted site visits and ethnographic observations in academic and non-governmental cartographic units, such as UN-affiliated mapping units and the Applied Research Institute Jerusalem (ARIJ), in order to better understand the community-relevant tasks and problems faced by cartographers working in a conflict region. Ethnographic observations were also conducted during events held by ICAHD: International Committee against House Demolitions, Jerusalem; tours at the City of David, Jerusalem, and the Regional Council of Unrecognized Villages (Negev); the Wadi Hilwah Information Center (Silwan/City of David, Jerusalem); the Yesha Council; Keren Kayemeth LeIsrael (KKL)/the Jewish National Fund (JNF); and Ir Amin (Jerusalem), among others. The combination of interviews, archival and library research, and ethnographic observations provides a rich set of data, which enabled us to interpret the textual and visual design of maps within the wider cultural and political context of their production, and to reflect on the history of cartographic constructions of Israel/Palestine through the words of the map-makers, planners, and surveyors as well as through the framings provided by political protagonists, policymakers, and scholars.

In This Book

In this book we trace how geography and cartography became entwined with politics, territorial claim making, and nation-building in Israel/Palestine. In particular we investigate the history of geography and cartography before and after the establishment of the state of Israel in 1948 and how maps produced by various

[5] All interviews cited were conducted by Christine Leuenberger. They were conducted between 2008 and 2015 in Israel and the West Bank. Interviewees' chosen represent an array of Israeli, Palestinian and International governmental and none-governmental organizations and academic experts that specialize in surveying, planning and map-making. All interviewees were promised anonymity thus interviews are coded by letter and year of production only.

governmental and non-governmental organizations before and after the war of 1967 solidify or question claims over territories and national identities. Before and after the 1967 war, maps of the region became one of the many battlefields in which political conflicts over land claims were waged. Aided by an increasingly user-defined mapping environment and new technologies and software that put mapping into the hands of the people, post-1967, Israeli and Palestinian interest groups and organizations increasingly used various cartographic techniques to demarcate disputed boundaries and inscribe particular national and ethnic identities onto the land.

To understand the role and function of maps in nation-state-building we trace the history of mapping in Israel/Palestine; we also analyze the visual and textual devices encoded in maps of the region and how they become visual stories that speak to the context of their production.[6] We probe what Clifford Geertz calls "cultural patterns" and "programs" (Geertz 1994) that can inform both the textual as well as the visual stories embedded within the maps examined here, and how they enhanced state- and nation-building. The project of state- and nation-building is an ideological endeavor that comprises not only the development of a state's infrastructures but also the national discourses that become "naturalized" and turned into "common sense" (Sutherland 2005). Various hegemonic discourses informed by religion and politics, as well as social cohesion, have helped co-produce the Israel nation-state. Maps and the narratives they tell have also become powerful tools to help create, maintain, and transform such widely shared hegemonic visions. For Palestinian stakeholders, maps have also increasingly served as tools to delineate and thereby attempt to consolidate a bounded national territory in the hope of establishing a fully recognized and sovereign State of Palestine in the future.

In particular, we show how various Israeli social and political groups and organizations produce maps that are grounded in particular sets of discourses or genres that have become part of the Israeli "common sense." This analysis of discourses and generic forms of mapping enables us to decipher the systematic differences and similarities across different types of maps (Günthner and Knoblauch 1995; Luckmann 1986; Wood and Fels 2008). We also trace generic features of Palestinian maps that are designed to depict a territory that is not yet under Palestinians' sovereign control. Informed by cultural blueprints, Israeli and Palestinian map-makers use graphical, symbolic, and linguistic tools to invoke authority, appeal to particular audiences, elaborate social concerns, and make political statements.

[6] While for cultural studies scholars, the circuit of culture would also include the study of consumption as a crucial stage in the production and reproduction of cultural processes, in this book we refrain from doing so. Numerous other studies on mental maps and in-situ interpretation of maps have pointed to fruitful ways of inquiry (Ben-Ze'ev 2012; Brown and Laurier 2005; Vertesi 2008).

Our interpretative analysis of the maps presented here focuses on the following elements: visual signifiers; textual signifiers; the demarcation of the space; and the social, cultural, political, and historical context of production. By visual signifiers we mean a map's projection and scale, the use and size of certain symbols (for example, thickness of lines and arrows), the level of cartographic detail, the choice of and contrast between various shades and colors, and the use of visual metaphors to emphasize or amplify a message. Textual signifiers include the naming of places, headings (their height, typeface, and color), written justification of methods, and other textual signifiers that can embed certain meanings within a map to reveal its purpose and target audience (Pickles 1992). We also examine the demarcation of spaces, which creates spatial order and hierarchies. This can be done by marking certain borders, routes, and sites; emphasizing or omitting different geographical and infrastructural features (Harley 1991); and selecting a certain level of detailed spatial coverage (for example, two- or three-dimensional spaces, topographical features) and spatial complexity. Most maps rely on a combination of visual and textual rhetorical devices. Depending on a map's function, purpose, context, and target audience, however, different rhetorical devices tend to be more or less dominant. For instance, maps that appeal to their scientificity for validating claims may include more textual signifiers than political campaign posters, which frequently use the demarking of space and prominent visual signifiers to appeal to voters.

Historically, both Israel and the Palestinians have used maps as geopolitical tools in their respective attempts to build their nation-states. Maps have many geopolitical functions: they serve to evoke a sense of national identity; they are essential for a state's ability to govern a territory; and they help define the extent of the nation-state in a contested region. The 1947 UN Resolution established a Jewish state side-by-side with a Palestinian state, which precipitated the establishment of the Israeli nation-state. The Palestinians, however, continue to insist on full recognition of their right to a sovereign and independent state. These struggles over land and its rightful owners in Israel/Palestine exemplify processes under way in other states across the world, whether in South Africa or Ukraine; they all are engaged in disputes over territorial boundaries, national identities, and the territorial integrity of nation-states. Maps, no less, have become crucial tools in these struggles.

The structure of the book is as follows: In chapter 2 we examine some of the ways that nation-states have been made—through narratives, ideas, and practices as well as through technologies and infrastructures—and how this has been reproduced in Israel/Palestine. Various disciplines were recruited to the service of nation-state-building. Cartography helped stake out a territory; history and archaeology were to make claims on it; statistics was to enumerate citizens; engineering was to build the state's infrastructure; environmental science was to transform the landscape; and geographers were called on to formulate a new geography of the new homeland that was in line with the challenges of forming a new society. At the same time, the Zionist vision and a widely shared hegemonic Jewish metaculture as well as the quasi-state institutions of the Yishuv (the Jewish community in Palestine before the establishment of the state of Israel) and its support from the Jewish diaspora contributed to the establishment of the Israeli state. Throughout the 20th century, the high-modernist state used science and technology to take on its people as a state project. Israel exemplifies how the use of science and technology contributed to the belief that societies, their people, and their territories could be known, managed, and improved. Science and technology charted grand new futures for societies, furthering scientific and technical frontiers as well as expanding the power of states, leaving behind all those people and lands that were not considered part of the state-building process.

Chapter 3 focuses on the importance of surveying and mapping for state-making in the 20th century and its importance for knowing and claiming the land in historic Palestine. We discuss how the Mandate of Palestine's Survey of Palestine (which was established to survey Mandatory Palestine) surveyed parts of the region. Their modernist ethos to register the land converged with Zionist visionaries to make it their own. With the Hagannah, the Jewish paramilitary organization during the British Mandate of Palestine, looting the Survey of Palestine, the Israeli state-in-the-making had access to cartographic material that helped them win the 1948 war (also referred to as the war of independence in Israel and the Nakba—the catastrophe—by Palestinians) and facilitated their statecraft. Post-1948, the Survey of Israel designed a new unified triangulation system, enabling the production of maps. The Israeli state also introduced a novel land tenure system. The seemingly imprecise land allocation practices common during the Ottoman Empire were pitted against a technocratic, modernist conception of land ownership that, by virtue of its implementation, dispossessed many Arab landholders. However, enforcement of technocratic regulations depends on humans. Indeed, the process of land registration reveals how surveyors who would go to villages to ascertain land rights were the human and, at times, weak link in doing so. Nevertheless, at the end of this process, 93% of land had become Israeli state land. The transformation in the land regime in Israel/Palestine thus attests to how new legal precepts in tandem with science and technology helped establish a modern, territorially defined state. While the Western scientific and legal paradigm enhanced the transfer of land, it also seemingly legitimized and depoliticized the new land regime, and made it seem part of the natural order of things and an inevitable outcome of modernity.

Chapter 4 investigates how Jewish organizations and Israeli institutions, before and after the establishment of Israel in 1948, produced various maps that helped foster an imagined national community and helped build the state. The Jewish

National Fund, in particular, become a powerful socializing agent into notions of territory. Its widely disseminated Blue Box, that was used to collect donations for buying land in the Holy Land Fn. The term Holy Land usually refers to territory roughly corresponding to modern Israel, the Palestinian Territories, western Jordan, and parts of both southern Lebanon and Syria. Jewish, Christian and Muslim religious traditions regard the territory as holy. Helped brand the territory and territorialize Jewish identity. Moreover, upon the establishment of a new state, re-naming territory is a pre-condition for the transfer of territorial control. After 1948, the newly appointed Governmental Names Committee established a Hebrew toponomy of the land. Yet top-down naming practices often encountered bottom-up resistance by local municipalities, as ideological directives would mix with local politics. Furthermore, for a new state, an atlas also becomes a powerful representation of its national story. The Israeli atlas reconstructs Israel's history, Zionist achievements, and the state's modern technical and scientific prowess. At the same time, maps of a small Israeli territory in the midst of the Arab world evoke the myth of Masada, as yet again, Israel seemingly represents the last Jewish stronghold under siege by enemy forces. Last, in pre-1967 Israel, the political landscape was ideologically split between various political parties, ranging from Mapai, Mapam, Maki, to Herut. Their different visions for a new society, a new human being, and a new state commanding a yet to be defined territory became part of cartographic debates that still reverberate in Israeli public discourse to this day.

Chapter 5 deals with post-1967 Israel/Palestine. We discuss how at the end of the 1967 war between Israel and its neighboring states, Israel had more than tripled the territory under its control, which intensified ideological rifts over Israel's territorial contours. As the rise of new mapping technologies and software turned cartography over to the people in the 1980s, maps became yet another way to swing the sword. While geographers and politicians increasingly found themselves embroiled in controversies over how to map Israel/Palestine, various political parties and organizations, ranging from the Likud Party; to the Yesha council (representing Jewish settlers within the oPt), to neo-Zionist organizations, such as MyIsrael, partook in map wars. They drew on various discourses associated with the Israeli right wing, such as the "secular expansionist" discourse (advocating territorial expansion for security and defense purposes); the "national religious" discourse (that was inspired by the political movement Gush Emunim encouraging Jewish settlements in Greater Israel) as well as the "Jewish rootedness in the land" discourse (that supports settlements as a historic and religious duty). Arguments for expanding Israeli territory for security purposes go hand in hand with the desire of the "national religious" and the "Jewish rootedness in the land" advocates to establish settlements in Greater Israel as a historical and religious duty. Their maps tend to rely on powerful visual, textual, and often emotive rhetoric to eliminate and marginalize Arab spaces, at times presenting them as a threat to Jewish presence in the region, and they tend to "Israelize" the territory and thus lay claim on it.

Chapter 6 examines how, post-1967, what are considered left-wing peace and human rights organizations (that are also referred to as the "peace camp") likewise used maps in order to put forth their geopolitical visions of an Israeli territory delimited by international law while also drawing on scientific cartographic conventions. Maps produced by the "peace camp" are informed by a range of very different discourses, which include concerns about Israel's occupation strategies, its compliance or non-compliance with international law, its demography, the need for the recognition of Palestinians' human rights and historical presence in the region, and the feasibility of particular territorial solutions. Organizations such as Peace Now, B'Tselem, the Geneva Initiative, and Zochrot use various visual and textual signifiers to communicate concerns in regard to territorial annexation, propose territorial compromises for possible peace negotiations, and challenge Hebrew topography by retracing alternative Arab topography in the search for historical justice. Such maps tend to invoke legal as well as scientific standards in order to instill the maps with authority and persuasive power in the attempt to increase the legitimacy of the geopolitical visions put forth.

In chapter 7 we explore the significance of borders in Israel/Palestine. The mapping of Israel's borders are where top-down mappings by colonial powers or clueless politicians intersect with complex regional realities. The history of border-making between Israel and Lebanon, Egypt, Syria, Jordan, and the West Bank all speak to what makes for "good" borders and better neighbors. The infamous Green Line (the 1949 armistice line, marking the internationally recognized boundary between Israel and the West Bank) exemplifies how a thoughtless delineation of the boundary by a bad map-reader with a thick pencil can reverberate across time and space for decades. Generally, delineations without regard to local conditions only fuel disputes over territory and can, in conjunction with ineffective national and bi-nation policies, negatively impact cross-border regions, economic development, and social interconnectivity across the border region. With many of Israel's boundaries in flux over the years, the Survey of Israel tends to emphasize not only the temporary status of boundaries but also to favor the representation of Israeli territorial claims. The stories of Israel's many boundaries reveal that there is no technocratic solution to boundary-making. Instead, stable boundaries were based on delineating them with the locale in mind; binational negotiations between policymakers and politicians; and binational teams of surveyors and experts for whom science could become a tool for establishing trust and therefore engage in better diplomacy.

Chapter 8 looks at how, throughout the 20th century, the rise of the Zionist national movement paralleled the strengthening of the Palestinian national movement. The struggle over Palestine also manifested itself in the history of surveying and mapping, and Palestinians' right to do so. After the Hagannah looted the Survey of Palestine, the Palestinians were left with few cartographic resources. The lack of maps of their own weakened their negotiating position during peace negotiations with Israel. Yet it was not until the 1993 Oslo Accords that Palestinians had a mandate to develop the territory under their jurisdiction. Their attempt to establish the State of Palestine went hand in hand with their effort to survey and map their territory. Consequently, in an effort to produce maps of their own, various governmental and non-governmental organizations produced maps for both building the nation and establishing a state. Logo maps of historical Palestine served to enhance national belonging; and cartographic reconstruction of pre-1948 Palestine retraced an Arab toponomy of the land. Concurrently, maps for building the State of Palestine delineate the territory in line with international law, strengthening Palestinians' case for territorial sovereignty. Such maps are also vital for governance, land allocation, and development. The lack of territorial sovereignty, restricted access to aerial photos at a suitable scale (due to Israeli restrictions), largely donor-funded mapping projects as well as the lack of a national mapping agency, encumber Palestinian mapping efforts to establish a state that could ascertain the rights of otherwise stateless people.

In chapter 9, we conclude that the story of the establishment of the Israeli nation-state exemplifies some of the main ingredients of nation-state-building in the 20th century. Israel came into being where historical narratives, national imaginations, scientific and technical know-how, human and material resources, and national and international support intersected. In the struggle to establish a nation-state, cartography had become crucial for both building the nation and for building the state. With the 1993 Oslo Interim Agreement, Palestinians also started to survey and map the territory allocated to a future State of Palestine, with the expectation that they would, within five years, have full sovereignty over the West Bank and Gaza. For Palestinians to survey, map, and plan for the territory is crucial for establishing the legitimacy and functionality of a future state. At the same time, their attempt to map their land—and the production of various alternative maps by various organizations—is challenging the top-down mappings of the Israeli state and its dominant geopolitics, which had produced many silences and omissions that can come back to haunt the visionaries of dominant geopolitics of Israel/Palestine. As boundaries continue to be controversial and the Israeli-Palestinian conflict remains intractable and unresolved, Israel/Palestine provides different governmental and non-governmental organizations, interest groups, and political protagonists ongoing fodder for persistent map wars. The focus on nationally based cartographic discourses in Israel/Palestine thus provides insights into the complexity, fissures, and frictions within internal political debates, but it also reveals the persistent power of the nation-state as a framework for forging identities, citizens, and alliances.

2

Imagining Nations and Making States

Some Prerequisites

> Few issues are as central to contemporary politics as that of co-operation between effective and legitimate states (OECD 2009).

Introduction: Imagining Nations and Making States

Under international law, states are commonly understood as entities that possess a permanent population, an assigned territory and a government. Independent, sovereign, stable, effective, resilient, and nationally and internationally legitimated nation-states are the bedrock of the international political system; and its stability depends on member states' ability to manage scarce resources, secure and protect citizens and their property, enable orderly economic exchange, assure transparency and accountability, and maintain legitimacy and resilience in the face of change or adversity.[1] In an era of globalization, functioning and strong states that conform to certain standards of conduct, transparency and accountability have become ever more important (Organization for Economic Cooperation and Development [OECD] 2009). For the OECD, supporting state- and nation-building are central for international relations and global stability; "nation-building" entails strategies a government uses to forge a strong sense of national belonging among the population, and "state-building" consists of creating national institutions to safeguard citizens, develop infrastructure, and enhance economic and political activities.

Consequently, a strong nation is one whose people share a common national identity as well as a functioning polity. Nation-states, like other cultural phenomena, can therefore be understood in terms of how nationalist ideas, symbols, and myths are enacted, maintained, and internalized by their people. However,

[1] States, however, have historically also been associated with violence, oppression, and repression as they may suppress various ethnic and religious identities, enforce taxation while providing minimum service to their citizens, and often fail to provide for political representation, institutional accountability, and the rule of law (OECD 2009).

The Politics of Maps. Christine Leuenberger and Izhak Schnell, Oxford University Press (2020).

DOI: 10.1093/oso/9780190076238.001.0001

not only ideas but also practices and materiality are enrolled in the engineering of states. To understand nation-states we therefore must not only attend to ideas, histories, and practices that make up a national culture, but we also need to scrutinize the "material strategies employed by governments for the control and 'improvement' of land, resources, and people" (Carroll-Burke 2002, 76). What is more, ideas are only effective if they inform practices and become institutionalized in material configurations (Carroll-Burke 2002). What then are some of the conceptual tools that address how ideas, practices, and material configurations make up nation-states? And how does Israel, in particular, exemplify some of these processes?

Imagining Nations and Boundaries

Benedict Anderson (1983) famously proposed that nations constitute "imagined communities." As he saw it, national belonging and identities are part of cognitive, imaginary, and discursive constructs that can be made and remade in different places and spaces. Accordingly, different national communities imagine themselves in different ways. There are many ways to foster a sense of national belonging and identity, and national maps have historically served as tools to do so. As they tend to emphasize borders, maps turn the nation's territorial shape into a recognizable logo, which can create the imagined community of a nation. Indeed, regardless of the time period in question, modern scientific maps have always had a nation as their subject and maps' origins lie in the very rise of the nation-state (Wood, Fels, and Krygier 2010). While, many times, national maps came to express an association of a community with a territory, maps sometimes also precede national sentiments. For instance, British maps delineated the territorial entity of India and thereby co-created Indian national identity. It was the map that connected previously disconnected and fractious regions and populations (Day 2008; Edney 1997). By creating the bounded "geobody" (Winichakul 1994) of India, surveyors facilitated the establishment of the British Empire, awakened Indian nationalism, and in 1947 established the border with Pakistan, which led to intense civil strife and the population transfer of Muslims and Hindis across the divide. Similarly, in Siam, Rhodesia, and Burma, maps preceded national sentiments (Collier 2006; Anderson 1983; Winichakul 1994). In the early 20th century, Zionist organizations also produced maps that helped territorialize Jewish identity, which was in the past constituted less by territory than by ethnicity, tradition, and religion (Leuenberger and Schnell 2010).

Maps not only express or at times create cultural homogeneity, but they also become tools to imagine the "other" who lives on the other side of the nation's boundaries, turning these borders into meaningful dividers between peoples.

For example, the establishment of the French state and its borders not only created a homogenous French polity but also gave rise to perceptions of cultural divergences with people who lived on the other side of the border (Hobsbawn 1992; see also Sterling 2009). The reinforcement and securitization of borders can, even in modern times, reinforce perceptions of otherness and notions of cultural incompatibilities across divides (Konrad 2011; Leuenberger 2006, 2011b; Vallet 2014). Maps and their material manifestation in territories can thereby help construct national belonging as well as cultural differences.

Maps alone, however, are empty signifiers (Barthes 1994). Their meanings, cultural significance, and set of underlying assumptions are only powerful if they can capture the imagination of a people. Since the introduction of compulsory schooling, education has become one of the most important socializing agents for a nation. Scholars have long pointed out that education crucially contributes to a citizenry developing national sentiments and encountering national ideologies and belief systems (Smith 1991; Gellner 1983; Lamm 2002; Hobsbawn 1992). Indeed, in post-1948 Israel, education became a fundamental part of forging a sense of national belonging.

Various disciplines were recruited to the service of nation-state building. In historic Palestine, and then Israel, Zionist historians published histories of the Israelites as commencing during the biblical period in Palestine and cumulating in the Zionists' redemption of what they consider the ancient homeland of the Jewish people. By emphasizing certain historical periods while deemphasizing or excluding others, such histories helped to legitimize the establishment of a Jewish state in Palestine. However, given the many centuries of Arab rule in Palestine, the Jewish state faced a "permanent external and internal legitimation crisis" to which, according to Baruch Kimmerling, "religion provided an answer" (1998, 51). The Old Testament served as "our charter for the land" (Kimmerling 1998, 51), which could establish a continuous link of the Jewish people to the land in Palestine. School curricula that focused on the First and Second Temple Period (also often described as the Jewish Golden Age in Palestine), and evaded and eliminated such contentious historical issues as Palestinian land claims, could strengthen the bonds between students and their territory and provide historical legitimacy to Jewish land claims (Bar-Gal 1993; Schnell 1999, 2000, 2002; I. Hoffman 2002; Miles 2011; Podeh 2002; Paasi 1999).

History and religion are not the only means to embed a people in a territory, but the geography of Israel, was also seen as a way to facilitate the alleged "primordial attachment of 'our' people to the land, thus strengthening even further the contemporary claims for territorial control and sovereignty" (Newman 2001, 239). Education in geography was designed to assist an immigrant population to "implant" themselves in the Land of Israel, helping them to develop a "sense of place" and an emotional connection and loyalty to the country (Bar-Gal and

Bar-Gal 2008). With the establishment of the Israeli state, scholars emphasized the need to educate a "new Jew" who knows the land, is attached to it, and will become rooted in his/her own homeland (Schnell 2001, 2004, 1999). Professional geographers were called upon to formulate a new geography of the new homeland that was in line with the challenges of forming a new society.

The first Israeli geographers had received their formal education in Germany and therefore were inculcated in the German tradition of *Heimatkunde* (homeland studies). Such homeland studies dovetailed with early Zionist efforts to instigate educational programs dedicated to pupils knowing their country. It was a way for promoting patriotism, delineating national boundaries, encouraging a sense of place and understanding of the cultural roots of the nation (Schnell 1999). Field trips and guided tours to ancient archaeological sites and biblical locations were extolled as a way to help new settlers become familiar with the land and its link to the Jewish past. They served to justify historical claims on the land and helped foster nationalism and attachment to the increasingly taken-for-granted homeland (Liebman and Don-Yiḥya 1983; Almog 1997).

During field trips to Holocaust monuments and war memorials, schoolchildren would also hear stories about those who sacrificed their lives for Judaism and for the nation. Inscribed on such memorials often are the words: "In their death they gave us life." It is such maxims that were to arouse an emotional attachment to the land and a patriotic commitment to the Jewish people, their history, and their future. At the same time, students would read geography textbooks that stressed what were deemed to be Zionist accomplishments: establishing settlements, conquering and irrigating the land, developing techniques for draining the swamps, and overcoming hostile environments so as to make the "desert bloom" (Alatout 2008; Frenkel, Shenhav and Herzog 1997). Such portrayals were based on biblical conceptions of Palestine as a land of abundance that had been subject to "neglect and destruction under Arab political and cultural regimes" (Alatout 2008, 960). Consequently, Jewish presence in the area was to return "the land into a living space bustling with agricultural activity" (Alatout 2008, 960). The discipline of Israeli geography thus emphasized the achievements of early Zionist settlers while also relaying a certain historical and environmental narrative about the land.

Besides formal geographical education in schools, various governmental and non-governmental organizations, such as the Israeli Defense Forces (IDF), the Ministry of Tourism, and the Society for the Protection of Nature in Israel, conducted educational activities. Their teachers and tour guides brought geography to the masses by organizing tours of relevant historical and geographical sites. With the rise of right-wing politics in Israel at the end of the 1970s, the Israeli Ministry of Education made a concerted effort to expand the study of subjects that were thought to enhance a sense of nationalism by increasingly supporting such informal study programs. The combination of such formal and informal educational efforts served to strengthen the relationship between individuals and the land and remained an important means of territorial socialization (Buttimer and Fahy 1999; Herb 1997; Bar-Gal and Bar-Gal 2008; Black 1997; Paasi 1999). As a result, in Israel/Palestine, more than in long established nation-states, space, territory as well as maps can provide conceptual tools through which identities are formed and strengthened (Newman 2008). Geographical knowledge therefore has been, and continues to be, part of "making" Jewish Israeli citizens and is inevitably tied to the nation-making enterprise.

Nation-building and citizens' identification with a territory, a society, and a culture can hardly be forced on a people. Indeed, the Berlin Wall was built in 1961 in order to stem the exodus of East Germans from the East German socialist experiment, only for it to fall in 1989 amid popular dissent. Long-term cultural power, therefore, derives not from force, repression, and coercion but from achieving popular consent (Gramsci 1971). In a consensus-driven, hegemonic society, different social actors and groups are less likely to engage in a *Kulturkampf* (a culture struggle). Indeed, one of the most effective tools of nation-building is the building of a dominant hegemonic political culture in which there is a

> high degree of consensus, a large measure of social stability . . . in which subordinate groups and classes appear to support and subscribe to values, ideas, objectives, cultural and political meanings, which bind them to, and "incorporate" them into, the prevailing structures of power. (Storey 2009, 80)

While history, religion, and education can help craft a sense of national belonging, nations are most effectively made and re-made with the consent of a people.[2]

When hegemonic consensus is enforced by a critical mass of intellectual, moral, political, economic, and cultural actors and gets expressed through an "ideological state apparatus" (such as education, the mass media, religion, and the culture industry), there are inevitably conflicts, resistances, and agents of opposition. These advocates of diverging visions may come from above and from below. A strong system of alliances among advocates and the state apparatus, however, may mobilize the majority and incorporate and co-opt opposing and

[2] Strategies for producing consent may vary. For instance, the success of Thatcherism in Britain stemmed from its articulation of a new "common sense" of how to organize markets and how to establish a new social order. The implication was that critics of dominant political positions were challenging taken-for-granted and common sense notions of what constitutes a seemingly natural social and moral order. This makes the notion of "common sense" one of the most powerful hegemonizing discourses available (Schwarzmantel 2009; Sutherland 2005).

resistant views. To do so can lead to a compromise equilibrium that does not threaten the common vision of a polity. Indeed, the weaker and more fragmented the opposition, the stronger is the state's ability to assert itself and succeed in the state-making process (Migdal 1989). The achievement of such a hegemonic consensus is not a given, rather, it arises at

> a very particular, historically specific, and temporary moment in the life of a society. It is rare for this degree of unity to be achieved, enabling a society to set itself a quite new historical agenda, under the leadership of a specific formation or constellation of social forces. Such periods of "settlement" are unlikely to persist forever. There is nothing automatic about them. They have to be actively constructed and positively maintained. (Hall 1986, 15)

A commonly shared hegemonic vision has therefore always been associated with major social transformations.

In Israel a "hegemonic metaculture" (Kimmerling 1998, 50) was a crucial ingredient in the establishment of the Israeli nation-state. Indeed, according to Stuart Hall (1986), historically,

> Hegemony is not exercised in the economic and administrative fields alone, but encompasses the critical domains of cultural, moral, ethical and intellectual leadership. It is only under those conditions that some long-term historic "project"—e.g., to modernize society, to raise the whole label of performance of society or transform the basis of national politics—can be effectively put on the historical agenda. (Hall, 1986, 17)

The forging together of various social, economic, and political forces also enabled the establishment of a hegemonic Jewish meta-culture in Israel. They include such mechanisms for nation-building as a widely shared sense of a unique Jewish history and collective identity, a long-standing apolitical and non-territorial cultural nationalism, a common language, and a religion that is enacted through an array of symbolic and ritualistic practices. Such commonly shared cultural capital paved the way to Jewish nationalism and a sense of shared purpose. For Baruch Kimmerling, religion, in particular, became a tool to mobilize a people to become part of the Zionist vision:

> The secular Zionist movement used and abused the Jewish religion and its symbols for mobilizing Jewish immigration to "Zion," and a tremendous effort of nation building and culture creation. Religion served as a legitimacy-generating mechanism both inside (the Jewish society) and outside (the non-Jewish "world"). (1998, 66)

As sociologists of religion have long noted, religion and its constituents—sacred items, transcendental entities, and supernatural phenomena—also remain empty signifiers, unless they are enacted and maintained within a community of believers (Durkheim 1965). Judaism entails the enactment of various sacred practices and rituals, whether they include the lighting of the Shabbat candles or the celebration of Passover. Besides commemorations and religious holidays such as Shabbat, the Jewish New Year, Yom Kippur, and Passover, secularized collectively held symbols and rituals such as the national anthem, the flag, the celebration of the "nation's birthday," Independence Day, and days of commemoration of the fallen in the wars and the victims of Holocaust (Kimmerling 1998; Migdal 1989) provide an array of festivities and ceremonies that are rooted in Jewish culture and tradition that create and sustain communities and provide a shared sense of purpose and identity.

In tandem with long-standing collectively shared traditions and a nationalist culture, post-1948 Israel was also characterized by very fuzzy boundaries between society and the state with certain organizations superseding that divide (Michael 1989). For instance, the early dominance of the Zionist Labor Party, which at different times was renamed Ahdut Ha'avoda (1919–1930), Mapai (1930–1968), and the Labour Party (1968–1977), provided mechanisms for disseminating particular political visions, including Zionist-socialist precepts, with ease to the entire Jewish community in both pre-1948 Palestine and post-1948 Israel (Peri 1983). According to Zionist-socialist sentiments, workers who had immigrated to the country to become pioneers were to become part of the power elite. Consequently, the mission to redeem the land, to reclaim Zion, and to establish a Jewish state by working the land, draining the swamps, and making the desert bloom, enforced the dominant national collective ethos and a sense of a common purpose (Frenkel, Shenhav, and Herzog 1997).

Besides commonly shared cultural practices and political visions, the pre-1948 Jewish community that had started to settle the region in the late 19th century had also established civil society organizations that became quasi-state institutions. The Yishuv constituted a state-in-the-making for the Jewish non-sovereign community in Mandatory Palestine, fostering political and economic development (Horowitz 1989). The Yishuv comprised a host of highly developed administrative, political and economic structures. For instance, the Histadrut (the General Federation of Workers) claimed a membership of 70% of Jewish workers in Palestine by 1926, offering a range of services including health care and education; it also became a de facto arm of the state and party policies (Michael 1989) and fostered a sense of nationalism by drawing on culturally shared myths derived from socialism, Jewish history, and national doctrines. The Yishuv's representative body Knesset Israel and its executive arm Va'ad Leumi (the National Council) included the various political parties in its governing body. Their culture of "bargaining and comprise" (Horowitz 1989: 30) served them well in building consensus in the push toward establishing a state. The Zionist movement also built the core of a paramilitary, including not only the Hagannah and its elite striking force the Palmach, but also youth groups trained to defend the Yishuv against its Arab neighbors.

The links between the Yishuv and the Jewish diaspora had also been a critical source of support, funds, and legitimacy for the Jewish nation-building project in Palestine. The Yishuv, while largely isolated from its Arab neighbors, was also economically, politically, and ideologically linked to a wide network of Jews and Jewish organizations around the world and tied into various international alliances. Such international alliances, as well as diplomatic and material support from the World Zionist organization and the Jewish Agency (which channeled international assets and funds into the Yishuv), proved to be crucial in the state-making process. Indeed, the diaspora served "as a source of manpower (immigrants), capital resources (the national funds), political support (influence on those who grant international recognition), value orientations (the immigrants' ideological commitment) and cultural traditions (a model of community organization)" (Horowitz 1989: 44). Consequently, the borderline between the Jewish community in Palestine and the diaspora was (and remains) permeable; in many ways, the Jewish diaspora represents an extension of the Jewish community in the Holy Land.[3]

It was also early attempts by the world's Zionist movement that ultimately provided Israel diplomatic recognition through the British Balfour Declaration of 1917, which consisted of a letter from British Foreign Secretary Arthur J. Balfour to Walter Rothschild, the leader of the British Jewish community, stating that "His Majesty's government view with favour the establishment in Palestine of a national home for the Jewish people, and will use their best endeavours to facilitate the achievement of this object" (Balfour 1917, 1). In 1937, the Peel Commission headed by Lord Peel argued that the Mandate had become ungovernable and recommended its partition into an Arab area that was to be united with Transjordan and a Jewish area. The Arab leadership was opposed to the petition at the time, while the Jewish leadership accepted it as it presented an opportunity for sovereignty. In 1947, the United Nations (UN) General Assembly passed UN Resolution 181 calling for the partition of Palestine into an Arab and a Jewish state with Jerusalem designated a "corpus separatum" (a separate entity) under international auspices. Consequently, Jewish sovereign control over

[3] The permeability between the World's Jewry and the Jewish community in Palestine/Israel also has profound effects on what constitutes Israel's polity, democracy, and concept of citizenship (Yiftachel 2006).

territory in historic Palestine was crucially advanced by international alliances supporting a Jewish homeland in Palestine (Galnoor 1995; Tarrow 2013). Moreover, both, Jewish migration to Palestine as well as the Yishuv's international alliances meant that trained cadres of bureaucrats and experts became part of the institution-building project. These cadres often "identified their own survival and mobility—their careers in society—with the success of the state as an autonomous organization" (Migdal 1989, 20). Establishing a functioning state therefore became intertwined with the personal and career interests and opportunities of various cadres, politicians, and officials, while at the same time assuring the political survival of the Yishuv.

The local and transnational institutions and organizations that were active during the time of the Yishuv, both within and outside the Jewish community in Palestine, constituted a state-in-the-making. Their institutional interlinkages not only provided political and economic capital, but their representatives were already united by a common ideology and religion as well as a commitment to the Jewish community in, and outside of, Palestine. Consequently, these "Jewish para-statal institutions" (Migdal 1989, 10) were characterized by a social cohesion, social control, and lack of fragmentation that most other states in formation do not possess. Their strength and reliance derived from a number of factors. The institutions in the Yishuv gained from the strength of Jewish migration to Palestine while the fragmentation of a heterogeneous Arab people and their relatively weak political and economic organizations affected the Arabs' level of social cohesion, social control, and ability to mobilize a population for a common cause and future. This helped create an environment ripe for colonization. Also, the British displayed a laissez-faire policy toward their colony in Palestine (Horowitz 1989). Their mixed policies toward its Jewish community while providing material resources, recognition, and a significant amount of autonomy to the Yishuv (which precipitated Zionists efforts to mobilize and strengthen their own leader and institutions) helped increase the capability of the para-statal organizations. Moreover, the fear of, and conflict with, the Arab community that was living side-by-side, but separate, from the Jewish community, served as a unifying force like no other.

The transnational organizations and institutions such as the Jewish Agency and the World Zionist Organization that united the Jewish community in Palestine with Jews around the world, as well as the Yishuv's organizational structures provided "important assets—bureaus, politically and bureaucratically experienced personnel, capital, symbols, legitimacy" (Migdal 1989, 10). Moreover, due to the strong commitment to a Zionist vision, the endurance of commonly shared and unifying cultural and religious symbols, and a common cultural ethnic consciousness and mission, Israel, more than other states-in-the-making, was to be more successful in dealing with opposing forces and dissent.

Indeed, the Israeli state was "much better at getting people to obey its rules once they are made" (Migdal 1989, 3). By the early 1940s, the Yishuv was a politically viable and virtually autonomous political system (Horowitz 1989) and it was these various organizations and institutions, material and human assets, international alliances, and commonly shared ideological visions that ultimately become the core of the Israeli state in 1948.

Such close interlinkages between an array of actors, institutions, and organizations that all shared common political and economic aims in the Yishuv in pre-1948 Palestine also came to reflect the close military-society nexus in post-1948 Israel. Across different cultural contexts, compulsory military service has always provided a powerful source of political and cultural socialization. In Israel, Jewish Israeli citizens have to serve while Arab Israelis are exempted from service. Service in the military molds soldiers' perceptions of citizenship and provides powerful mechanisms for legitimizing a hegemonic and militarized social order. This is particularly true in a society that is perceived to be under constant threat from outside enemies. The military ethos, too, therefore encouraged national sentiments and helped build the nation (Ben-Eliezer 1998; Kimmerling 1983a; Gal Levy and Sasson-Levy 2008). Army officers also frequently become co-opted into the political elite, which encouraged close formal and informal ties between them. Consequently, there were few confrontations or conflicts of interests between the military and civilian sectors (Lissak 1983).

Moreover, the long-standing Israeli-Palestinian conflict meant that Israel's constant need for self-legitimation, and the sentiment that the Israeli state is under persistent threat of destruction, led to a constant and extensive mobilization of human and material resources. Consequently, the mobilization of the military and civilians alike is part of the "rules of the game" in Israel (Kimmerling 1983a). External threats, whether perceived or actual (Bar-Tal and Antebi 1992), combined with a mission to prove the legitimacy of the Israeli nation-state thus unifies and integrates a polity and provides, more than elsewhere, the conditions for a mobilized and recruited citizenry that will act in the interest of the commonly shared state-building mission. In the first three decades of its existence, and during the crucial stage of nation-state building, Israel was therefore a "highly mobilized and collectivistic society" (Kimmerling 1998, 67) that enforced a cultural hegemony in the interest of a shared vision for a new Hebrew culture and a nation-state in the making. Israel thus did not need a Berlin Wall to force its people to strive toward a Zionist and socialist vision of a new state; rather, cultural, social, political, and economic forces intersected for a people to be willing pioneers and advocates for a new Jewish society in the heart of the Middle East.

Making States

States, however, cannot only be imagined; they also have to be engineered. While a citizenry needs to be mobilized, states are "socio-technical systems" (Carroll-Burke 2002, 104) consisting of not only humans, but also of non-human things, such as buildings, lands, and material infrastructures. Indeed, state formation is not only political. "Otherwise the Palestinians, Kurds, and other state-less 'nations' might be expected to imagine themselves into statehood" (Carroll-Burke 2002, 78). Instead, a range of scientific knowledge and practices are required to transform a land into state territory. Indeed,

> governments mobilize science to act upon, and thereby engineer a *country*, a country that is composed fundamentally of territory *and* population. The state system is a "socio-technical" system, a "skein" of networks that includes persons, technologies, conventions, protocols, papers, buildings, and so on. (Carroll-Burke 2002, 79)

There are various material practices that have historically been crucial for building states. The counting and defining of citizens through population census; the building of physical infrastructures such as roads, canals, public buildings, and industries; the transformation of landscapes and local ecologies, including planting forests and rerouting rivers; the building of monuments and museums; and the political appropriation of archaeological discoveries to link a people to a territory are just a few of the ways nation-states have been fostered into being. Moreover, to survey, map, and name the land and develop commonly shared standards, codes, and measurements by which to do so have also significantly contributed to the material infrastructure of a state. In the following we will discuss how such practices have facilitated state-building and we will focus on how they have helped constitute the Israeli state in particular.

Each state has to first "make up" its citizens. Ian Hacking (1990) points out that the rise of the nation-state in the 19th century produced an "avalanche of printed numbers" (p. 2). National governments started to classify, count, and tabulate their citizens to an ever more precise and descriptive degree, marking and identifying their various traits, preferences, and affiliations. Indeed, the quantification of a citizenry constructs the notion that everyone is accounted for in the census and has a clear place within it. A national census thereby not only helps imagine a nation but also helps define, construct, and govern its citizens. Indeed, one of the tasks of state-building includes conducting a census, as it provides a technology through which to enumerate a citizenry within their bounded territory (Carroll 2006). Also, colonial states, have always used a census in order to conduct a "demographic topography," which organized an ethnic-racially divided population according to "educational, juridical, public-health, police, and immigration bureaucracies" (Anderson 1983, 169).

The state of Israel also came into being with the help of the newly established Israeli Central Bureau of Statistics (Leibler 2004). Its first major project was a population census that eliminated certain categories of people from the bio-body of the Israeli state (such as many Palestinians who were not in residence when the census took place), and deemed others, such as Bedouins and Mizrakhim (Jews from Muslim countries) enumerable, yet separate from Ashkenazi Jewish Israeli citizens. To enumerate people thus went hand in hand with differentiating them in terms of racial categories (Bowker and Star 1999). The Israeli census thereby defined the parameters of the governable and enumerable population, as their respective characteristics became categorizable and legible to the state. A census as well as ethnic categories (and corresponding rights to citizenship and ownership) can thereby serve to not only include but also to exclude certain kinds of people as part of the national bio-politics (Yiftachel 2009a). The coming into being of such classificatory infrastructures—that make it possible for a people to be defined, counted, and classified as either part of or as excludable from a geo-national body—is crucial for the making of a state.

Besides the need to define and count a people, states also build up their physical infrastructure. Lands, roads, buildings, sanitary infrastructure, rivers, canals, and harbors that are connected to political and economic infrastructures are central to the material state. Indeed the power of a state lies in its ability to engineer and coordinate a physical environment as well as its penchant to govern, regulate, and maintain it. The power of governance and state control is thus embedded in such large-scale infrastructures. The building-up of material infrastructures is also vital for colonizing a land. In 17th-century Ireland, estate owner William Petty started to survey the territory, draw maps, and improve roads that become a model for future planners eager to engineer Ireland (Hacking 1990). Reclaiming and cultivating land as well as building drainage systems and road networks became part of the colonial process there. Such infrastructural improvements had the ideological connotation of civilizing a land and a people. Roads served multiple purposes: they enabled industry to develop alongside it; they facilitated the control and policing of formerly inaccessible areas that were thought to harbor outlaws; and they served the military as they eased interchange between Dublin and the military barracks in Glen Crie (Carroll-Burke 2002). Similarly, in South Africa, the railway system helped form the skeleton upon which the modern South African nation was built as it fostered economic independence and promoted tourism, settlements, and investments (Foster 2003).

In Israel, large-scale infrastructure also became a tool to build the state. The first large-scale public project—the drainage of the Hachula swamps and the construction of a national water canal—came to symbolize the power of the state to overcome nature through organization, modernization, and technological innovation (Schnell and Rosenberg 2011). In the 1950s, roadways were built to foster economic activities and connect the national frontier to the center of Israel so as to enable the efficient maneuvering of military forces to Israel's boundaries. The more recently constructed Highway 6, the trans-Israel Highway that runs from the north to the south of the country, is the largest Israeli infrastructural project ever undertaken (Garb 2004). It is linked to the rest of Israel's road network and connects the periphery to the center. It also fulfills various geopolitical aims: it strengthens Jewish developments along the Green Line and provides a corridor for troop and military equipment.

While states build up their physical infrastructure, they also imprint their culture onto the territory under their control. Landscapes thus do not represent nature but become culture; at the same time, defining what is "natural" can itself become an ideological tool. To be sure, claiming certain natural elements—whether they are flowers, fruits, birds, or topographies—as being indigenous to a particular place while equating them with a national collectivity is "one of the most affective as well as effective ways in which belonging are claimed" (Billig in Elliott Weiss 2010, 206). The transformation and appropriation of landscapes has thus historically been a crucial tool for projecting a state's power. For instance, the French landscape gardens from the early modern period represented "orderly domination of the earth through engineering" (Mukerji 2002, 4). A walk through the great gardens of Versailles reveals how irregular spaces could be transformed into ordered spaces that attest to the nations' capacity to control the landscape, including its water supply, horticulture, and architecture. The garden—as an engineered environment—hereby served social and political purposes and was used to legitimize the modern French state. Similarly, the emergence of scientific forestry in Prussia and Saxony between 1765 and 1800 were also linked to state-making initiatives of the period (J. C. Scott 1998). Forests were conceived and planted so as to be counted, manipulated, measured, and assessed for purposes of planning. They were to be part of a controlled environment that could be surveyed, supervised, harvested, and sold.

In Israel, the Jewish National Fund conducted afforestation projects that fulfilled various symbolic and practical functions for building the Israeli state. Tree plantings came to signify the return of the Jewish people to what was deemed to be their historical homeland; newly planted trees were to make the land habitable and create a new landscape. After the establishment of the state, this new landscape became the foundation for agricultural innovations and rural planning. Afforestation served to defend cultivable areas and protect roads and railways against sand movement. Forestation hereby became a technique to redeem the land, and tree plantings came to signify what the Israeli state had declared to be state land, to which Palestinians or Bedouins had allegedly lost their claims (Young 1993). As laws stipulated that state lands could not be sold once transferred to the state, Jewish land holdings increased while Arab-Palestinians became structurally excluded from the possibility of owning or acquiring what had been declared Jewish lands (Yiftachel 2006). The planting of trees thus enforced Jewish ownership over the land by "strengthening the grasp on Palestine" (Liphschitz and Biger 2004, 304) and thus became part of the war between Israelis and Palestinians (Braverman 2014).

Besides the enculturation of landscapes into a new state system, the construction of monuments and museums also serve to foster national cohesion and cultural unity by co-narrating a national history. Earlier in this chapter we mentioned how school field trips to such monuments together with stories about Zionist accomplishments are a powerful means to socialize children into a shared geographical and historical narrative of Israel. Also for the general public, memorials become sites that can evoke the past in terms of various national and political myths and ideals (Young 1993). Such state-sponsored memorials serve political purposes. For instance, monuments of the graves of anonymous soldiers are powerful symbols affirming the legitimacy of the nation-state. Memorials inscribed with such sayings as "In their death they commanded us life" connote soldiers' loyalty and commitment to the nation as well as their willingness to sacrifice their lives for the sake of an imagined community. The soldiers' namelessness only amplifies the message that those in whose name they died are obligated to honor their sacrifice and continue the valuable fight. The Israeli landscape is dotted with such monuments and memorials that praise soldiers' and pioneers' sacrifices for the collective, and they can elicit in onlookers a sense of loyalty to the soldiers who purportedly scarified themselves for Israeli independence and prosperity (M. Azaryahu 1996; M. Azaryahu and Golan 2001).

Museums, too, can serve to construct a national history that justifies the existence of the nation-state. The Jerusalem-based museum and memorial, Yad Vashem, was conceived with the founding of the state of Israel—it was to commemorate the Holocaust, provide Jews murdered during the Nazi era posthumous Israeli citizenship, and to represent the Israeli state's justification and self-definition as a refuge for the global Jewry. It is the state's official historical memory of the Holocaust and has become part of Israel's civic religion. Foreign dignitaries are brought to Yad Vashem before meeting the state's leaders. Such visits serve as a ceremonial preamble to state talks and reveal Israel's civic religion, which has also tied the horrors of the Holocaust to the right of the world's Jewry to establish a state in Palestine.

While histories can be forged above ground, through memorials and museums, they can also be made below ground. Archaeology can confer a tangible material meaning on a historical tale. The excavation of archaeological items can link selected historical periods to the present and confer upon them a significance in the chain of historical events that make up the chosen history of a nation. The neglect, removal, and abandonment or the exposure, reconstruction, and preservation of certain archaeological sites and strata can produce one particular history of a land while concealing alternative historical narratives. In Israel, archaeological activities have historically served two main aims: the exposure of the Jewish past during the First and Second Temple Periods and the tracing of early Christianity (Greenberg 2009). This archaeological focus was particularly widespread until the late 1960s when such archaeological finds also served the new state's need to legitimize its existence by pointing to early Jewish history in the region. The archaeologist turned national hero, Professor Yigael Yadin from Hebrew University, headed some of the biggest national projects: the search for the earliest available traces of Judaism in the Judean desert as well as the reconstruction of Masada. Masada became a central national symbol, signifying Jewish heroism and devotion to the national territory. The story tells us that it was a fortress built by early Jewish settlers, which after a Jewish revolt in the region in the first century CE, was besieged by Roman forces. Rather than succumbing to Roman forces, however, the rebels committed mass suicide, so as not to live in servitude (Ben-Yehuda 1995). Archaeologists like Yadin, despite a lack of archaeological evidence, provided the archaeological foundation for the history of Masada, and the Zionist movement claimed this heroic tale as it served as a model of its ideal new type of Jewish identity:

> a Jew whose identity was grounded in a glorious history, in ancient warrior's pride, who saw him or herself connected to the land and to the legacy of ancient heroic figures that—like him or her—were willing to live and die for the land. (Ben-Yehuda 2009, 338)

To this day, organized tours and field trips bring schoolchildren and soldiers to the fortress of Masada to impress upon them this tragic, yet heroic, story.

After the Israeli de facto annexation of the West Bank in 1967,[4] Jewish settlers in particular continued to propagate an archaeology that made deliberate efforts to reveal connections between Jewish history and what was considered to be the biblical homeland. As Judeo-Christian sites were excavated with the Bible as a charter to the archaeology of the region, Muslim-Arab roots received comparatively little attention in archaeological research. One of the main reasons for the relative neglect of Muslim archaeological sites was the Antiquity Laws

[4] For further detail on the legal status of the oPt, see Kretzmer 2012.

implemented by the Israeli state. They defined which time periods would constitute antiquity and specified the types of objects and monuments that were to be excluded from the provisions of the ordinance (El-Haj 2002; Greenberg 2009). Despite the contentious nature of what constitutes an archaeological site, the Israeli Ministry of Education endorses annual field trips for schoolchildren to archaeological excavations into the oPt that focus on the biblical and messianic past (Schnell 2012). Hereby, such activities and government-sponsored programs define the West Bank as part of the national territory of Israel and contribute to the construction of the Israeli nation-state as the rightful home of the Jewish people.

Right-wing NGOs are also actively "making" archaeology. While many Israeli archaeologists argue that excavated Jewish sites, such as the Kingdom of David and Solomon, were smaller and less developed than expected based on biblical stories (Finkelstein et al. 2007), organizations such as the right-wing Jewish association El-Ad (also known as the Ir David Foundation, that funds Jewish settlements across East Jerusalem) insists on the value and cultural sophistication of the City of David's archaeological site in East Jerusalem. According to Israeli archaeologist Yonathan Mizrachi, however, the El-Ad-sponsored digs exclusively focus on the Jewish tradition and ignore other strata of Jerusalem's multi-cultural history (McGirk 2010; K. Adler 2010). The City of David has thus become a highly contentious site, not least because it is situated in the middle of the Palestinian-populated neighborhood of Silwan. For Mizrachi (2011), archaeology has thus become a means to "judaize" the land as it denies Palestinians their land by other means. Archaeological excavations hereby become a way of collecting facts and producing knowledge through which the "historic Jewish nation-home" emerges visibly and materially from the material remnants of the past, while erasing "the question of 'Palestine' from the history of the Israeli state and society, which has become, quite simply, the nation-state of and for the Jewish people" (El-Haj 2002, 55).

High Modernist States and the Construction of Legible Citizens

As we have seen in this chapter, it is at the intersection of a nation's imaginary discourses and material cultures that nation-states are conceived, maintained, and stabilized. The fundamental role of the symbolic realm of culture and its intermarriage with a nation's material infrastructure are truly modern phenomenon. As James C. Scott points out (1998) pre-modern states knew relatively little about the societies, people, and resources over which they presided. Sources of information were fragmented. Local currencies and measures were regionally varied and local societies were opaque to outsiders. Indeed, for ancient regimes, the main incentives to make citizens legible were to tax them, control them and keep order, and to draft them into the army.

The high-modernist state, however, uses science and technology to take on its people as a state project. Its ideology requires novel levels of interventions into society to make the formerly illegible legible. It is the rise of democracy that requires new forms of legibility and a "bird's-eye view" of a state's subjects so they become legible to its gaze and fine-grain identification (J. C. Scott, Tehranian, and Mathias 2002). In other words, as modern states are removed from the governed, they have to nevertheless be able to govern "at a distance." Human activities thus become inscribed in documents and legible, codifiable, and comparable through statistics and record-keeping.[5]

As the purposes of the modern state are broadening, statecraft also requires an ever-greater knowledge of the social and physical environment that is to be governed. Since the beginning of the 20th century, governments have expended much effort to mobilize the sciences, whether they be the "soft" or the "hard" sciences, to engineer their countries and their citizens. Geography was to stake out a territory, archaeology was to make claims on it, statistics was to enumerate people in line with bureaucratic logics, anthropology was to understand indigenous "others" and make them governable, and the environmental and information sciences could be used for strategic military purposes. Indeed, throughout the 20th century, the widespread belief was that societies, their people, and their territories could be known, managed, and improved. Science and technology charted grand new futures for societies, furthering scientific and technical frontiers as well as expanding the power of states. Indeed, scientific and technical knowledge had become and continues to be a fundamental block of state building (Eghigian, Killen, and Leuenberger 2007).

Policymakers and the international community of nations are increasingly concerned about the proliferation of "weak" and "fragile" states (OECD 2009). To imagine a nation into being and to engineer its state requires resources, support, and coordination. The nation-state of Israel came into being where historical narratives, national imaginations, scientific and technical know-how, human and material resources, and national and international support intersected. The Israeli state-building process was thus, and continues to

[5] The accumulated knowledge about citizens and lands can be used for beneficial and detrimental purposes. For instance, the naming and identifying of Jews in central Europe during the 19th century provided them with full citizenship yet was later used to deport and exterminate them. Such atrocities could only be enforced by a high modernist state that had made the social landscape legible by making a people identifiable by name, address, ethnicity, and religion.

be, a modernist project. How this nation-state was first imagined, narrated, surveyed, and mapped, and how its maps turned into tools used by many to struggle over the making of the Israeli national consciousness, narrative, and territory, and how they also became a vehicle for governance, is the story we want to tell here.

3
Map-Making and State-Building

Introduction: Maps for War and for Peace

The historian Eric Hobsbawn (1992) once pointed out that the 20th century was the century of ideologies, yet it was also the era in which science and technology came to facilitate statecraft like never before. Science and engineering were called upon to assist in state projects from road building to nuclear engineering, and the social and human sciences could assist in managing populations, repatriating soldiers, and conducting psychological warfare (Herman 1995; Gerovitch 2001; Hacking 1990). One characteristic of World War I was "a new and more intimate relationship between science, the state and the military" (Heffernan 1996, 505). Science and technology crucially supported war efforts, benefiting the sciences in the long run in terms of funding, research opportunities, legitimacy, and prominence. Some scientists joined the ranks of foot soldiers, and geography, too, gained its fighting medals. Indeed, already in 1907, the then president of the Royal Geographical Society (RGS), George Taubman Goldie, declared, "War has been one of the greatest geographers" (1996, 504).

Also, during peace times, however, history has shown that to construct a nation-state, we need guides to its territory, such as maps. Maps that transform unwieldy territories into knowable, visible, and governable units have been one of the most important tools of governance and control (Andrews 1999; van der Woude 2008; Black 1997; Carroll 2006; Day 2008; Harvey 2009; J. C. Scott, Tehranian, and Mathias 2002). The modern state presupposes a uniform property regime that is legible, controllable, and governable from the center. As a principal legal repository of property delineations, the cadastral map is one of the fundamental building blocks of state-making. The modern state took it upon itself to delineate and demarcate rights to land, to register and map land ownership, and to ensure tax revenues. Cadastral maps were the tools for this job—they verified rights for the lands and delineated plots, including both publicly and privately owned lands. For this purpose, land is divided into registration block, with each village or town divided into several registration blocks, which are divided into parcels. It is the parcel that is the basic unit of the cadastral network. Such cadastral maps consolidate the power of those who commissioned them. The fact that they become the nexus of particular configurations of power (Gavish

The Politics of Maps. Christine Leuenberger and Izhak Schnell, Oxford University Press (2020).

DOI: 10.1093/oso/9780190076238.001.0001

and Kark 1993; J. C. Scott 1998)[1] becomes clear when we examine the role of maps for colonial conquest.

Map-making has in the past been tied to conquest and has served the aims of colonizers. The British surveyed and mapped territories ranging from Ireland, India, and Egypt to Palestine to facilitate the governing of the British Empire. Surveying and mapping enabled them to better manage the colonies, modernize infrastructures, re-parcel land according to what were considered to be the most rational and scientific means, and exercise their control over the land. To this day, maps serve to assert territorial claims.[2]

Such mapping practices have always been for the benefit of some and to the detriment of others. In colonial contexts, maps depicted new settlements and traced indigenous populations in order to control, conquer, or ultimately dispossess them. It became an instrument for the literate and was legible to administrators but often indecipherable to illiterate and rural populations. For state officials, map-making has been part of knowing the land (Day 2008) and was a precondition for supplanting societies and asserting claims to land. Indeed, to map is to know and to control, while the failure to survey and map a territory also weakens claims on it. Given the importance of surveying and mapping to conquest, colonization, and state-building, cartography also become a crucial tool for the Israeli state's attempt to wage war and make peace throughout the mid-20th century.

Mapping the State: Finalizing the Triangulation Network

With the establishment of the Israeli state in 1948, mapping in the service of state-making become essential to govern the land and to centralize control over the entire territory. At the time, Joseph Elster, the first head of the Survey of Israel, specified the Survey's three main task: delineating and demarking boundaries; producing a cadastral survey of land ownership; and transforming the surveys into official maps (Szancer 2001). The Survey was also to regulate surveying and mapping activities and institutions in Israel. Faced with the post-war

[1] For example, in post-socialist Transylvania, newly implemented notions of territorial spaces and property rights reversed the previous socialist rules for property holding, which enabled the state to more effectively act "at a distance" (Verdery 1994).

[2] For instance, when the French mapped Martinique in the late 18th century they recorded sugar, cotton, and coffee colonies and provided information vital for possible future hostilities with Britain (Black 1997). In 1992, after the breakup of Yugoslavia, maps were used also as tools to argue for its lands to either be part of, or separate from, historical Greece, which ultimately facilitated the creation of the Republic of Macedonia. With Russia's occupation and annexation of Crimea in 2014, their maps were remade so as to portray Crimea as part of Russia while US maps of the region continue to represent Crimea as occupied territory (Chappell 2014).

task of building a state and developing and securing a country, the military and civilian surveying and mapping units worked together. Indeed, the need for this continued integration was spelled out by the Survey of Israel:

> From the start of activity of the Survey of Israel in 1948 . . . the mapping work was dictated by the immediate demands that arose from the circumstances of the establishment of the state during a war. The department had to provide maps for the military, to update the existing maps, to complete and to map areas that have not yet been mapped and to change the maps into Hebrew. (Srebro, Adler, and Gavish 2009, 24)

The first governmental bulletin in 1950 specified the responsibilities allocated by the Israeli government to the Survey of Israel; it was to publish maps based on systematic measurements that describe the country's topography as well as the boundaries of land parcels. Previous measurements and surveys of lands had been sporadic, diverse, and lacking in common standards. As a result, they could further development or land ownership rights in specific locales only. Such practices could not provide a comprehensive and unifying spatial system of control over a national territory; only a unified network that enabled centralized control over the territory through rational and technological means could accomplish this.

The British Mandate government had already established the triangulation methods, a basic control network that assigned fixed points on the ground in triangles and measured the angles among them subjecting territorial spaces to a mathematical-geometric formula; however, the triangulation network remained unfinished by 1948.[3] The British survey started the triangulation with a series of triangles extending between Urim and Zemach, excluding the Negev and Judea desert. At the time, surveyors did not perceive the Negev as part of the Land of Israel, nor did they see any urgency in surveying and mapping what to them was an unpopulated desert area. For the post-1948 Survey of Israel, however, developing more accurate maps of the Negev was a matter of urgency.

During the 1948 war, the Jordanian Army controlled the Arava valley, which provided the shortest route to Eilat and Akaba. To get around them, military and mapping units that were to be sent to Eilat had to find traversable routes through the mountains and valleys of the Negev. Through cooperation between

[3] Use of the triangulation method had become standard to British surveying and mapping systems around the empire. Their motivation for such mapping efforts were, on the one hand, militarily strategic and on the other hand, part of a push to establish land rights and thereby increase their tax base (Ferguson 2002).

the military and the civilian mapping units, maps were created that integrated information from a number of sources, albeit with different scales and levels of accuracy. Among the sources were British maps at a scale[4] of 1:25,000 (1 cm on the map corresponds to 250 meters on the ground), Jordanian maps scaled at 1:100,000 (1 cm on the map corresponds to 1 km on the ground) (made in 1943–1945), British military maps of the area of Akaba from 1945 at a scale of 1:25,000, and various British air photos. The eventual military success—and the fact that the Negev was ultimately recognized as part of the State of Israel—is to large extent due to the success of mapping out these frontiers in the shadow of war.

But these maps were quickly obsolete. After the 1948 war, the ceasefire lines with Egypt and Jordan meant that Israeli control of the Negev was not assured. Israeli Prime Minister Ben Gurion stressed at that time the importance of the port of Akaba as a gateway to the Red Sea and to Asia and Africa. By the time Joseph Elster took over as the first director of the Survey of Israel, he stressed the inadequacy of the Negev maps and emphasized the importance of accurate surveys as the basis for remapping and controlling the Negev. In a letter from April 1950 to the Minister of Labor he writes that surveying the Negev is a necessary precondition for mapping it, and that the basic geodetic data from existing Negev maps are unreliable and that they do not reach the detail and standards necessary for military maps (Szancer 2001). The Survey of Israel subsequently designed a new triangulation map that could serve the purpose of creating topographical and cadastral maps of the new territories that the Survey of Palestine had not included (as shown in Figures 3.1 and 3.2).

These maps make clear the differences in the coverage of the triangulation networks and how they changed from the era of the British Mandate to the updated Israeli triangulation network, which included the southern Negev and the Judean desert. With the introduction of new and seemingly more accurate surveying technologies, from aerial photography to satellite images and geographic information systems (GIS), the Survey of Israel continued to adjust and standardize the new triangulation network, cumulating in a national network in 1997 that utilized 160 base points. The new unified system eventually secured continuity across boundaries, which was advantageous for military mapping and also enhanced coordination with neighboring states, in terms of trans-border geodetic scientific research and development projects conducted with Egypt and Jordan.

[4] Map scale refers to the relationship between distance on a map and the corresponding distance on the ground. Large scale maps makes features look larger, showing a smaller geographic area; and small scale maps make features look smaller, showing a larger geographic area.

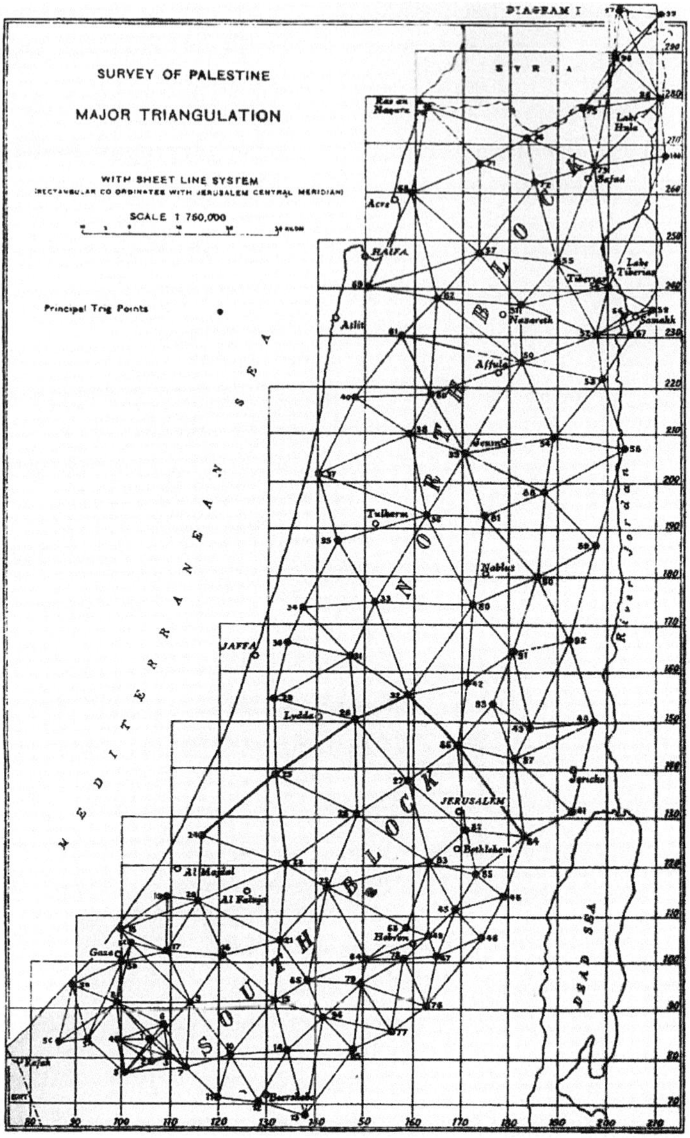

Figure 3.1. Survey of Palestine: Major Triangulation 1:750,000 "Triangulation Control Network at the end of the Mandate", in Srebro Haim, Ron Adler, and Dov Gavish, *60 Years of Surveying and Mapping Israel 1948–2088*. Tel Aviv: Survey of Israel, 27.

Figure 3.2. The updated Israeli geodetic Control Network including the Southern Negev and the Judean desert. In Srebro Haim, Ron Adler, and Dov Gavish, *60 Years of Surveying and Mapping Israel 1948–2008.* Tel Aviv: Survey of Israel, 29.

Such spatial systems replaced the traditional notion that conceived of landholders as the basic reference specifying their holding in respect to their neighbors. The transformation of the control system from focusing on humans to focusing on space is one of the more fundamental principles of modernist control systems. Such modernist control networks fulfill many functions: they provide the spatial foundation for securing land rights and regulating property taxes; fostering development, including building settlements and infrastructure, such as roads; and producing maps for military purposes and for delineating national boundaries (R. Adler and Gavish 1999). While surveying and mapping efforts started to build the unified network that spanned the region, the Israeli state also introduced a novel land tenure system. It is to the history of land and land tenure in the region that we now turn—how land was registered, measured, mapped, as well as bought and sold. We trace land regimes under Ottoman rule, the British Mandate, and the Israeli state.

The Ottoman Heritage and Its Land Regime

The Ottoman Empire's Land Regime between 1858–1914 crucially shaped land management in the region far beyond the demise of the empire. The Ottomans, in line with ancient civilizations of the Levant, adopted the basic proposition that land belongs to God, and as political leaders representing God on earth they were the de facto landowners; the Ottoman sultan owned most of the lands and distributed the rights to use them to his subjects. Nevertheless, the Ottoman government made various attempts to reform its system of land laws and registration. Their main goal was to distribute land ownership to the peasants so as to modernize cultivation methods, and register landownership for the purposes of extracting taxes as these were one of the empire's main sources of income.

The Ottoman Land Code (OLC), introduced in 1858, become the cornerstone of the land system in Palestine/Israel until the Israeli Land Law in 1969. The OLC specified various categories of land. Mulk entitled full ownership of land and was rare; such land holdings were usually small parts of land that were in close vicinity of a village. Miri land was land people could register in their name after 10 years of cultivating it; however, these land holdings ultimately belonged to the state; and Mewat (dead) land was uninhabited land away from the village and uncultivated, yet it could be transformed to Miri land if cultivated for longer than 10 years. Musha land was distributed every three years by the village elders among the peasants according to the number of males in the family. After three years the lands were redistributed so those who had more fertile land would get less fertile land and vice versa. On the other hand, peasants who did not cultivate lands for more than three years in a row turned their land into Machlul lands and thereby lost their right to that land. Wakf lands were owned by the sultan and secured for religious uses. Matruka lands belonged to villagers collectively and included public usage of land for roads and pastures.

One goal of the OLC was to register the land, but it did not implement a precise system for doing so. Even if land was registered, the description of the land often did not reflect the reality of the land. The Ottomans did not conduct a land survey, and land registration was done through the supply of deeds; however, these did not include exact geographical location of properties. Indeed, exactness was not of primary concern. Methods to allocate land rights were embedded within socially cohesive communities at the time and within techniques that adequately defined land rights for these communities. The need for registering land and having ownership documents seemed counter to village culture:

> Such villages were usually small communities with a high degree of cohesiveness and familiar long-term relationships between community members where unofficial social arrangements for land possession, the terms of which were clearly understood by the participants, developed as alternatives to the official system of registration. (Kedar 2001, 934)

While the Ottoman land system seemed disorganized and unclear from the outside (Kedar 2001), for the peasants themselves the system worked. For instance, under the OLC ordinance, Mewat land could be defined as a distance from the village that "a loud human voice cannot make itself heard at the nearest inhabited place to it." (Kedar 2001, 954) What mattered most for peasants at the time was whether the land was situated within hearing distance, which would also define the boundary of a settlement, and this made the voice as a unit of measurement more important than distances measured in terms of miles (Kedar 2001).

Furthermore, for villagers, informal agreements regarding land ownership seemed preferable, as official registration entailed being taxed and possibly drafted into the Ottoman army. Therefore, "many peasants refrained from registering land transactions, fearing that such registration would attract the attention of tax collectors to be used as a mechanism for military conscription" (Kedar 2001, 933). As a result, during the Ottoman period almost no land was formally registered, and only about 5% of the land in Palestine had been registered by the end of the Ottoman period.

In fact, peasants the world over had ways of resisting a state's advances into their territory which, at the time, seemed to work in their favor; in the long-run, however, it became their ruin as they did not have the necessary title deeds proving land ownership required by a bureaucratized system of governance embodied within the modern state (J. C. Scott 1987). The same applied to peasants after the establishment of the Israeli state. While some peasants decided to register their land and consequently received land ownership documents, others who avoided registration may have not been drafted into the Ottoman army, but they frequently lost their rights to land once the Israeli state came into being.

The Establishment of the British Mandate for Palestine and Its Land Regime

There had been small-scale attempts by various individuals and institutions from England, France, and Germany (as well as other countries) to survey and map greater Palestine. British emissaries and organizations had been engaged in sporadic surveying and mapping efforts of the region since the 1840s. Such attempts to survey and map the territory were multiple and manifold. They were conducted by an array of individuals and organizations for biblical, archaeological, geological, cultural-linguistic, or strategic purposes. These maps lacked common guidelines, standards, or measurement, as they were neither drawn according to uniform systems of coordinates nor in accord with a shared cartographic method. They also lacked any legal status.

By the first half of the 19th century, however, Britain needed an efficient transport route to its most important colony: India. With France controlling Algeria, Egypt occupying Palestine and Syria, and Russia posing a threat to the north, British surveyors started to map Palestine to be able to control land and sea routes (including the Suez Canal). Such gateways were to enable the efficient transport of military troops and goods between Britain and India (Goren 2002). Once the British government decided that Palestine had military and geostrategic value, the British War Office supported the Palestine Exploration Fund that had been founded in London in 1865 in order to conduct more comprehensive surveys and map Western Palestine between 1871 and 1877. It was therefore strategic and military considerations that led to the completion of the Western Palestine Survey (Gavish and Kark 1993).

By 1915, the Royal Geographic Society was also asked to prepare a map at a scale of 1:1,000,000 of the Ottoman Empire. With few accurate maps or surveys of the region that could be used, the RGS relied on observations and recordings of explorers, scholars, and intelligence officers who had worked there before World War I. Alarmed by a German plan to build a railway to Baghdad which was seen to threaten the British government's imperial route to India, it was of interest to the British intelligence service for the Palestine Exploration Fund to design an archaeological map which would help improve British intelligence about the region. Indeed T. E. Lawrence wrote at the time that the mapping project was "only meant as red herrings to give an archeological color to a political job" (Heffernan 1996, 513). After the Allied victory by 1918, RGS's 1:1,000,000-sheet map (in which 1cm on the map corresponds to 10km on the ground) was adopted as the base map for redrawing political boundaries in Europe and the Middle East. With Britain supplying a map of such detail and scale, the British had a strategic advantage. Even though RGS's cartographers provided maps, they were, however, for the most part, not called upon to address policy matters. Nevertheless, geography's contribution to the war and peace efforts at the beginning of the 20th century had proven the discipline's usefulness in state matters.

Only with the establishment of the British Mandate of Palestine (1922–1948) were the land regulations and registration systems modernized. The imperial modernist ethos, coupled with the demand for formal and systematic control over territorial spaces, meant the British government was "dissatisfied with the lack of formal order in the land" (Kedar 1996, 296). Indeed, they saw disorder everywhere. What worked for traditional villagers in Palestine was a thorn in the eye of the all-viewing state. The British government attempted to replace the widespread informal arrangements in Palestine with formal, state-enforced order. For instance, the Ottoman regulation that allowed villagers to gain rights to lands through cultivation for more than 10 years was abandoned. The people who continued to cultivate such land were instead treated as trespassers (Kedar 1996; 2001).

In an attempt to modernize Palestine's land system, the Survey of Palestine performed a cadastral survey and the Land Registry was established to deal with land settlement issues. At the time,

> no law had yet been passed in Palestine laying down standards of measurement and making the surveying by authorized surveyors obligatory. Nor had any standard linear or area measures been agreed upon. (Gavish and Kark 1993, 78)

During the 1920s, the British government initiated the systematic surveying, mapping, and registering of land as a basis for land reform and the enlargement of the tax revenue.

> The survey started with the production of village maps with fiscal blocks to facilitate the taxation of rural property. These maps then served as a basis upon which registration blocks were parceled and surveyed. (Gavish and Kark 1993, 79)

Yet the surveyors in the field were to be the quiet heroes and/or villains of the story. Given the fuzzy territorial boundaries between different types of land (such as what constitutes Miri lands), land allocation rights were decided less by formal guidelines, measurements, and surveying techniques and more by negotiations with the peasants. Operating in such a wide gray zone of inadequate information, the surveyors often had the power to decide between different types of lands. They decided whether the peasants' ownership claims were credible, where the boundaries between Miri and Mewat lands were to be drawn, and whether possession by cultivation of lands should be recognized. As Dov Gavish (2005) points out, during a survey in 1923 such weighty decisions were made on the day of the surveyors' arrival in the villages, and if villagers could not adequately define the boundaries of their plot, their land was lost.

While early surveying efforts focused on defining Miri and Musha lands to encourage land reform in Arab villages, mapping efforts increasingly focused on Jewish settlements. According to Dov Gavish and Ruth Kark (1993), starting in 1929 this was due to unrest in the Arab communities, culminating in the 1936 Arab revolt and the outbreak of World War II, which hindered the project's advancement. During this period, British surveying and mapping efforts shifted to the lands of the Jewish settlements. Arguably, surveyors preferred to work in safer territories, such as the plains and the valleys where their efforts overlapped with the Jewish settlers' desire to establish their legal ownership of the land. Indeed, Zionist organizations had encouraged British efforts to survey the land and settle land titles to facilitate Jewish land purchases and settlements in the spirit of the Balfour Declaration (Kedar 2001).

As a result, Jewish settlers not only generally welcomed the land surveyors and settlement officers into their communities, but they also appear to have done some of the work for them (Solel 1991). Jewish settlers closely surveyed and mapped their communities and kept detailed Land Books that were seen as some of the more reliable and comprehensive maps available (Gavish and Kark 1993; Gavish 2005). Arguably the British Mandate's motivation to survey and map the "settlement areas" of Palestine, rather than the more arid Arab areas of the Eastern Galilee and the Negev, can partly be explained by security concerns. Moreover, the Mandate's modernist ethos for formal order, development, and bureaucratic logic and its desire to get the job done converged with the Zionist project to survey and map the territory (Kedar 2001; Abu-Sitta 2007; see also Fields 2017).

Toward the end of the Mandate, the cadastral survey and land settlement fell behind schedule. Only about 5,000 sq km of the 26,000 sq km west of the Jordan River had been surveyed, much of which was among the most fertile land along the coastal plains and in the Jordan valley (Gavish 2005). These areas constituted the northern part of what was to become the State of Israel. As we discussed previously, the Negev, as well as mountainous areas (such as Judea and Samaria/the West Bank), were excluded from the survey as they were perceived to be less (or not) suitable for cultivation. By the end of the British Mandate in 1948 only about 20% of the total area of Palestine had been mapped and surveyed, which has fueled land disputes to this day (Gavish and Kark 1993).

A Map Affair: Where British Planning Converged with Zionist Visionaries

The British Empire's desire to modernize Palestine went hand in hand with Zionists' desire to make it their own. Not only did the British mapping of Jewish settlements enhance Zionists' ability to ascertain land rights and buy up land, but surveying and mapping the country brought a modernist order into the "Oriental East," and, as a result, made it more controllable and governable. Therefore

> Although the British Mandate has its independent agenda, the unintended consequences was more of a convergence, even if temporary, with Zionist bureaucratized conduct than with the Arab community. (Ben-Ze'ev 2011, 45)

After all, the Yishuv was "literate, bureaucratized, and organized within a tight net of national networks" (Ben Ze'ev 2011, 45). Moreover, many Jewish immigrants came from Eastern Europe and had brought with them a modernist ethos that they shared with the British imperialists. Palestine's Arabs, on the other hand, were a more rural and semiliterate population. The local fellahin were

> well acquainted with the features of landscape—they knew them from their daily practice of working the land. . . . However, their familiarity was often restricted to their own locality, extending to a radius of 10–20 kilometers or the nearest commercial town. For the *fellahin*, other regions often remained terra incognita. (Ben-Ze'ev 2011, 47)

Given their orientation toward their local environment, they were also a more divided population who did not identify with one umbrella national institution. They found themselves caught up in the imperialists' drive to modernize and develop the country: "The difference in the scale and spectrum of institutionalization between the Arabs and the Jews had a decisive effect on the two sides' preparation for the [1948] war and on its outcomes" (Ben-Ze'ev 2011, 45).

When it came to map-making, the British and the Zionists' aims seemed to converge even though their ultimate goals diverged. Jewish cartographers and surveyors worked side by side with the British Survey Department. Many had immigrated from Europe, where their training had incorporated the most-up-to-date geographical and cartographic techniques and expertise at the time, and where they also had established networks with Europe's cartographic communities and publishing houses. Such international interlinkages were crucial as, for instance, Austrian printing presses could print base maps of Palestine. As a result, all the senior experts in the Survey of Palestine were either Jewish or British. Indeed according to a Israeli cartographer:

> The survey of Palestine is one of the very few institutions of the British mandatory administration, which had very, I would say, excellent relations with Jewish scholars so that they were directly involved in the surveying and in the production of maps. (Interview B 2010)

Following the Arab Revolt in 1936, the British army worked together with the Hagannah to gather intelligence and topographical information and map Arab villages. A small office in a Tel Aviv apartment served as the Hagannah's base for its mapping unit. Pinhas Yoeli, an immigrant who had been trained in the latest cartographic approaches in Germany, headed Hagannah's intelligence and mapping units. As map literacy was seen as an important military asset, he taught map reading and navigational skills in areas with varied topographical features. Hillel Birger, Hagannah's topographic trainer at the time, argued that it was crucial that in the event of an armed struggle, soldiers should have access to Hebrew maps; he therefore requested that the Survey of Palestine produce maps in Hebrew.

By the 1940s, the Hagannah was collecting topographical maps for military purposes. They prepared "village files" that contained military intelligence and aerial photographs, and they also purchased, copied, or stole British maps that were enlarged for the purpose of planning possible operations and confrontations with Arab armies. At that time, the Hagannah improvised secret flights during which they took aerial photographs of Arab villages. As they had to take off with a civilian aircraft from a British airport, agents, with woman and baby in tow, explained to the British officers on duty that the "baby had asthma and the doctor said that if he will fly high, the air will do him good" (Ben-Ze'ev 2011, 52). When asked where they would be flying to, they explained it doesn't matter as long as it was high enough. They then proceeded to fly over Arab villages to take aerial photographs and map out the region. Upon return, an unsuspecting British officer remarked, "I suggest you take another pilot. Your pilot doesn't know where he's going. He just flies like that" (Ben-Ze'ev 2011, 52).

It was World War II, however, that furthered the convergence of British and Jewish interest in the region. With the war raging, the Survey of Palestine was conducting military mapping tasks, including mapping for "control, intelligence, navigation, and planning" (Srebro, Adler, and Gavish 2009, 22). For such purposes, topographical maps often provide the base map on which the military maps were then drawn. Indeed, most survey offices produced just military maps, or both military and civilian maps. As a result,

> signs and colors were erased. A new grid system was added, a gazetteer of place names was compiled, and there was special emphasis on whether roads

> were passable for military vehicles. . . . [I]t was these militarized maps that the Hagannah and the Palmach adopted and used for their own purposes in the 1940s, and especially in 1948. (Ben-Ze'ev 2011, 49)

As these maps become vital military tools, the British also enrolled over 30,000 Jewish soldiers and about 9,000 Arabs into the British war effort. The civilian surveyors who served under the British officers were all Jewish. According to surveyor Asher Solel, the relationship between the British officers and the Jewish ranks was very good. He pointed out that "we learnt a profession from them" (Ben-Ze'ev 2011, 50). By the 1940s the British army feared a German invasion of Palestine and "they were now more willing than before to cooperate with the Hagannah, which in turn was eager to acquire their training" (Ben-Ze'ev 2011, 48). The Jewish soldiers working with the experienced British army honed a range of military and cartographic skills. And when in 1941, the Hagannah created the Palmach, a "strike force" that was to become the Hagannah's elite fighting force, the British agreed to train 150 Palmach fighters in sabotage and scouting. "The Palmach offices collected information to update maps, prepared sketches, and wrote accounts on the location of police stations, the headmen of Arab villages, cisterns, and flour mills" (Ben-Ze'ev 2011, 48).

By 1944, the British government commenced a survey of Arab villages. Aims included addressing inequities between Jewish settlements and Arab villages, modernizing and developing rural Arab areas, tightening control over the region, and building up tax revenue. More than 100 villages had been surveyed by 1947, when the British Mandate was dissolved. With the dissolution of the Mandate, however, "the maps drawn for this effort did not yield their intended outcome. Instead, some were used by the newly formed Israeli army in the 1948 war to obliterate the villages that were meant to benefit from development" (Ben-Ze'ev 2011, 44).

After the war ended and the British Mandate was ending, the British planned to transfer some cartographic material that pertained to a future Arab state from the Survey of Palestine to a British military camp in Ramlah. From there it was scheduled to be transferred to Cyprus. The Hagannah was determined to stop that transfer. In April 1948, just before the scheduled transfer, the Hagannah broke into the Survey of Palestine. As an Israeli cartographer explained in an interview:

> Members of the Jewish underground penetrated the building at night . . . and stole . . . 70–80% of the material that was in there, so that there was not much left for them [the British] to take to Cyprus. The building itself—the Survey of Israel is the same building in which the Survey of Palestine was housed . . . so . . . the

> Hagannah could easily take control of the building and empty it to a very large extent. . . . [T]hey took the original drawings, they took away the printing plates, nearly all the maps and the survey material, so that they could easily within a few weeks after the British left, produce all the maps. They overprint[ed] Hebrew over the British maps. (Interview B 2011)

While protagonists disagree over the role of the British in this operation, with some arguing that the operation was done without the involvement and knowledge of the British, one interviewee recounts that he met the director of the Survey of Palestine years later who told him "that in fact he cooperated with the emptying of all the resources . . . and he was quite satisfied that it fell into Jewish hands and not into Arab hands" (Interview B 2011). Indeed he recounts that "he was friendly with the Jewish community" (Interview B 2011) and whether out of professional or political consideration he cooperated with the looting of the Survey of Palestine.

As a result, within a few weeks of the departure of the British, the building in which the Survey of Palestine was housed became the Survey of Israel. Although the British administration had not formally intended to transfer the department to the Jewish Yishuv, they in fact took it over in 1948. The Hagannah not only inherited the equipment and plates to produce the maps but also the maps themselves, which could easily be overwritten with Hebrew lettering; and they had also honed their cartographic and surveying skills through the many years of working with the British imperialists. As a result:

> When the state of Israel came into being, its cartographic life was very much based on the products of the Survey of Palestine. . . . Professor Yoeli . . . took over from the British—the Survey of Palestine and he produced all those maps in Hebrew for the Israeli army, the administration and so on—but he and others [had] . . . produced maps during the time of the British administration—very good maps based on the British survey and on other information that they collected—so that the state was very well supplied with maps on the day on which it came into existence. (Interview B 2010)

Therefore, upon the founding of the Israeli state, the Survey of Israel was fully operational and had all the necessary means to use cartography for building the state.

At the same time, while the British had planned to transfer some cartographic material to the Palestinian Arabs, after the Hagannah successful emptied the

Survey of Palestine, there was not much left for them. They had access to material in the Survey of Jordan (that had also been run by the British) which "at that time had some equipment and some maps of part of Palestine, but they were not as well equipped and updated [in comparison] to what the Israelis took" (Interview B 2011). The Palestinian Arabs thus lost out in the war of maps. While the Survey of Palestine had employed some Arabs, it was only a limited number, and they were mostly field surveyors. Also, the Palestinians did not possess a centralized organization, such as the Hagannah, that encouraged them to collect and produce maps. During the 1948 war, their local knowledge might have provided some advantages, but the maps of the Jewish forces nevertheless helped conquer the battlefield. While the Arab liberation army, such as the Egyptian forces, also used maps as they entered Palestine, the maps used could not compete with the large body of cartographic knowledge, skill, equipment, and material that had been accumulated by the Hagannah through training, theft, and secret spotting. Arguably, the Hagannah's commitment to the Zionist vision fueled their efforts to survey and map the territory and access mapping sources from the British Mandate government. Also, it was the British and Zionist commitment to science and technology, to making legible a seemingly illegible landscape through surveying and mapping, and in the process replacing diverse local knowledges, and implementing a modernist state, which ultimately transferred the reins from the empire into the hands of the Zionists in the name of progress and control.

The Israelization of the Lands: Transforming Arab Land to Israeli Land

After 1948 the state of Israel continued where the British left off. A functioning state needed a governable territory under its control. To achieve this, officials used the power of the law, the power of science and technology, and the allure of modernity to institutionalize and justify the dispossession of the Palestinian Arabs and the Israelization of the land. During 1948–1960, a series of new laws established new "ethnonational geographies of power" (Forman and Kedar 2004, 809), and the allure of objective science and exact technologies in the service of modernizing a seemingly chaotic and disorderly system provided the legitimacy to do so. Surveyors and mappers played a role in this process of Israelizing the land and dispossessing the Palestinian population in the early years of the Israeli state. It is their stories we now turn to.

Official Histories of Land Management

After 1948, surveyors crucially facilitated the process of registering land, determining ownership, and completing the triangulation system that the British survey department had started, but did not finish. A book published by Srebro, Adler, and Gavish (2009), *60 Years of Surveying and Mapping Israel 1948–2008*, to commemorate the 60th anniversary of the establishment of the state, portrays how the protagonists of the Survey of Israel understood their work to be pioneering and as establishing the preconditions for the successful operation of the state apparatus:

> The Surveyors are the pioneers before the camp. They are the first to be in the field, to get to know the field, and to map it. . . . When they arrive in the field, the field is usually bare, often unapproachable and lacks the basic supporting facilities. In their surveys, their maps and the databases that they produce about the space, the surveyors create the basic foundations for planning, for preparing infrastructure and utilities for paving roads, for building houses and for registering land rights necessary for maintaining an organized way of life. . . . By being a man with his feet rooted in the ground and his gaze looking forward the surveyor reflects the qualities of a leader. (Srebro, Adler, and Gavish 2009, 7)

This description is in line with the dominant Zionist myth; the land is described as desolate or bare, awaiting the people of Israel to redeem it.[5] Surveying and mapping was seen to have transformed a bare and unapproachable territory into a state territory by rational, systemic, technical, and modern means (Almog 1997). The surveyors are celebrated as pioneers, conquering the land through surveys, measurements, and topographical analyses, and opening up the frontier for development. Israeli educational textbooks similarly explain the Zionists' success in terms of their use of superior technologies and level of bureaucratic organization. "The pioneers" thus were understood to have redeemed the land, and the technical tools by which to do so came to justify the means (Horowitz and Lissak 1989; Schnell 2012).

In the Survey of Israel's 50th anniversary book, Ron Adler, director of the Survey of Israel from 1971–1992, also emphasized Israel's technological progress since the establishment of the state. He points out that the Survey of Israel upheld the British Survey Unit's commitment to using the latest technologies to survey

[5] Davis and Burke (2011) point out that such notions as "arid" and "desolate" lands are part of the colonial understanding of the environmental history of the Middle Eastern region that has been distorted by orientalist assumptions. Yet such conceptions of the Middle East went hand in hand with Jewish perceptions of their need to "make the desert bloom" and "redeem their land" from desolation (see also Davis 2016).

and map Israel (R. Adler and Gavish 1999). While the Jewish agencies in Palestine had perceived the British as the enemy who hindered land purchases, they were considered partners when it came to sharing survey technology. For protagonists from the Survey of Israel such cutting edge technologies were considered essential for development and they ascertained the leading role the agency plays in using the most advanced technologies (Srebro, Adler, and Gavish 2009).

While the power of technology can help claim a land, it is not enough. Scientific rationality and mathematical logic was also to capture complex physical and human realities. For Ron Adler and Dov Gavish (1999), geodetic-mathematically designed national maps reflected and mirrored the "face of the land"—including its natural and cultural landscape, land resources, and their legal and economic impact on the nation-state. Maps therefore were thought to provide an objective analysis of national space as they seemingly enable spatial thinking and spatial order which enables the identification and understanding of agency-environment relations (R. Adler and Gavish 1999).

With science and technology presented as crucial to the process of redeeming the land and building a nation-state by demarcating, developing, and defending it, professional histories by organizations such as the Survey of Israel tend to present the surveyor as the often forgotten hero by pointing out that "when the area is already developed and teeming with life the surveyor's crucial part is often forgotten" (Srebro, Adler, and Gavish 2009, 7). Even though unrecognized, the surveyors therefore nevertheless identify as pioneering experts who use science and technology to establish the hegemonic Zionist vision by redeeming the land for the Jewish people. As a result, they can be understood as part of the "recruited elite" of Israeli society, that during the first 30 years of the state, played a crucial role in the state-building process (Kimmerling 2001; Horowitz and Lissak 1989).

Besides their self-understanding as largely unrecognized pioneers who translated the Zionist vision into physical reality, Israeli surveyors (just like surveyors for the Survey of Palestine years earlier) were also the people who had to negotiate with the villagers in the field, mediating between Ottoman laws, popular habits, and new Israeli laws, in order to delineate the parcels and allocate land rights. During field visits, villagers were invited to present evidence of land ownership on a given day, and it was the surveyors who were the ultimate arbiters, deciding on the validity of the evidence provided. While villagers had the option of appealing to the Supreme Court, many did not, due to their lack of resources, procedural knowledge of the legal process, and non-recognition of the new state's justice system. As a result, the power to delineate the boundaries between private land and state land ultimately fell to a large extent to the surveyors.

It was during such visits that peasants often resisted surveyors' attempts to survey and register their land in units foreign to them. At times, they might also have not declared, have under-declared, or have stated false information concerning the extent of their cultivated acreage or harvest in order to avoid or to reduce their tax burden. Although many of these forms of resistance are shy of a political movement, according to James Scott, such forms of everyday resistance are "the peasantry's most common and durable weapon" (J. C. Scott 1987, 421). Lacking the power to influence legislation, they resist and quietly evade regulations and stipulations as the surveyors and mappers on the ground attempt to enforce them.

Then again, surveyors who operate on a provincial or national level may, at times, also accommodate to resistance, not least because they frequently depend on local networks of patronage. Everyday resistance can thus become an institutionalized practice that is quietly tolerated. While sometimes such everyday forms of defiance of certain legislative processes can become a means to change laws, in most cases there are no written records and no traces in the archives of such episodes (Le Roy Ladurie in J. C. Scott 1987). Nonetheless, villagers' resistance to the top-down authority of the nascent Israeli state to evade registering land could backfire in the long-run, as their land claims often remained unrecognized and some continue to be disputed in Israeli courts to this day (Kestler-D'Amours 2012).

Once the surveyors had measured out the territory and determined the state's right to certain lands, delineating it become an urgent priority for the Survey of Israel and the Division for the Registration and Settlement of Lands. As land rights had to be established before Palestinian Arabs could instigate land claims based on their cultivation practices under Ottoman laws, transforming land into state land, in the field and on paper, became a national priority as it was essential for the development of the state. Indeed, according to protagonists at the Survey of Israel:

> We see that since the early years of the state high priority has been given to the settlement of state-owned or state-managed land rights. It is likely that this priority was decided upon in Israel as a result of the large masses of lands for which the state claimed ownership and the need to delineate and demarcate those rights and to formally register them in order to enable their development according to national targets. (Solel 1988, 160)

For the Survey of Israel, the land settlement project was thus a procedural problem to be solved in order to foster much needed development:

> The immigration waves since World War II . . . required the development of a wide-range of building in the urban area and in the rural areas. . . . [T]he Survey of Israel was responsible during all those years for keeping up with the land settlements . . . updating transformations that were carried out in

> the cadaster . . . [and] preparing the plans for registration. (Srebro, Adler, and Gavish 2009, 36)

Some of the alleged "technical difficulties" included the "accurate delineation of the relevant parcels" (R. Adler and Gavish 1999, 17), which was necessary for land settlement and urgent development projects. Their urgency stemmed also from the arrival of over 1 million Jewish refugees in Israel in less than a decade. In 1954, a compensation agreement with Germany fostered investments in building basic state infrastructure. In the early 1950s, new towns and villages were planned and built to absorb the immigrants and agricultural advancements were activated to support a growing population. A new land regime that enabled the transfer of Arab land into Jewish hands was thus crucial for the many towns, villages, and settlements that sprang up all over the country.

All these areas had to be surveyed, delineated, and registered as state lands and parceled for planning.[6] In addition to surveying for settlements, surveyors were also called on to plot land to establish a transportation network connecting the new towns and villages to the general road and railway network, the ports of Haifa and Ashdod, and the burgeoning industries. Furthermore, the surveying and mapping work of the Survey of Israel facilitated economic development projects. Between 1954 and 1955 surveys were conducted for water projects, copper and oil fields, mines, and airport improvements in Lod as well as for mapping lands for afforestation (Szancer 2001).

For surveyors from the Survey of Israel, the process of land registration was an essential part of state-building, and land reform was essential for facilitating the Jewish development project. However, the politics of land acquisition and the associated dispossession of Palestinian Arabs remain a blind spot in the official histories of the Israeli state-making process. A former employee from the Survey of Israel, Asher Solel (1991), who was involved in the cadastral mapping project, points out that:

> the Ottoman land regime remained valid with minor changes during the British Mandate and within the state of Israel up to 1969 until the announcement of the new land law, when the state of Israel gained the rights over most of the lands that were not settled. (p. 159)

Solel recounts a number of types of land, such as "abandoned" lands that seemingly inevitably became state lands. He points out that based on the Ottoman, Mandatory, and Israeli land laws that pertained to the lands of "absentees," the

[6] To give a sense of the scale of the effort: in 1949 the government built 26,413 dwelling units for Jewish refugees, and in 1950 the number reached 45,000 dwelling units (Szancer 2001).

Israeli government acquired up to 90% of the state's territory. "Abandoned" and "absentee" lands refer to lands left behind by Palestinians during the period known as the Nakba during which more than half the pre-1948 Palestinian Arab population fled or were expelled from their homes and land between 1947 and1949, including during the 1948 war.

However, rather, than focusing on the discontinuity and rupture caused by the war and its accompanying displacement of people from the land, Solel establishes continuity between the different legal regimes (see also Kedar 2001; Forman and Kedar 2004). He presents Israel's transformation of Arab land to Jewish land as the self-evident consequence of the Ottoman and British land regimes. For him, new Israeli land laws seemingly established order in the previous disorderly land regime. As a result, the new Israeli land regime is presented as essentially apolitical, in line with Israeli law, and an inevitable outcome of previous land regimes.

Carmela Szancer (2001), another former employee in the Survey of Israel, addresses Arab land acquisition projects more explicitly. She points out that after 1953, the Survey of Israel and the Land Settlement and Registration Management Office both prioritized land settled by Arabs until 1948, for transformation into state lands and for Jewish settlement in order to avoid reclamation of that land by the Arabs. Consequently, surveys focused on 42 Arab villages in Galilee where the state expected to be able to legally claim 400,000 dunams.[7] Efforts also focused on the Negev, where it was assumed that at least in its southern parts no Arab claims would hinder the land acquisition project—an assumption that turned out to be too optimistic given the continuing land disputes between the government and the Bedouins there. As the Bedouin landholdings (mainly in the northern part of the Negev) were not registered under the Ottoman or British regimes due to their informal systems of ownership allocation, they largely lost their rights to the land under Israeli land law (Yiftachel 2009b).

The Establishment of the Israeli Land Regime

Sandy Kedar (1996) points out that historically, certain well-established practices and mechanisms have, in different cultural contexts, led to transformations in territorial control. Land can change hands through expropriation during ethnic conflicts, as it did in 1923 between Turkey and Greece, and in 1947 during the formation of the two new states of India and Pakistan. In addition, populations that have suffered ethnic discrimination such as the Tamil in post-1948 Sri Lanka or the Native Americans in the United States have been subject to practices of

[7] During the early part of the British Mandate a dunam was 919.3 square meters. By 1928, the metric dunam of 1,000 square meters was adopted.

land expropriation. However, one of the more common expropriation practices entails the use of law as a means of transforming or negating land rights. For instance, European colonials have long enacted legal mechanisms to seize, expropriate, reallocate, and reclassify land (Forman and Kedar 2004). The Israeli case exemplifies elements of all these forms of expropriation. The new Israeli state was established in the wake of British colonial rule; refugees of different ethnonational makeup either exited or entered the new state's territory; and the Israeli state developed legal means to "internally displace" the Palestinian Arab population within its borders and "judaize" the country (Yiftachel 2006).

Jewish colonization had began under the Ottoman Empire and intensified under British rule. Land was acquired through the open market as well as through the Jewish National Fund (JNF) and other land-purchasing companies. By the end of the 1948 war, however, only 13.5% of Israeli territory was Jewish owned, yet Israel was under Jewish sovereignty (Kark 1995). Palestinian Arab lands were limited to some 160,000 Arab villagers that remained within the Israeli state's territory. The rest was either Mewat or Miri land, the ownership over which was deeply contentious. The Israeli state used various legal means by which Arab land could become Jewish land (Kretzmer 2012; Kimmerling 1983a).

After the 1948 war, tens of thousands of Palestinian Arabs were classified as "absentees." As we have discussed earlier, the Israeli state seized land (left behind by Palestinian Arab refugees) by classifying land first as "abandoned land" and then as "absentee land." By 1953, the Knesset had enacted the Land Acquisition Statute, which legalized land transfer retroactively and transferred land to the Development Authority (DA). "Absentee" property could then be sold to the JNF or transferred to the DA, which transformed the land into state property, also known as Israel Land. Israel Land constituted all land that belonged to the state, the JNF, or the DA respectively. Furthermore, given the lack of land registration during the Mandate period, the Israeli government expanded its conception of what constituted state land. Legal land disputes often came down to what constituted Mewat and Miri land:

> The state strove to qualify the lands as Mewat, thereby establishing ownership, whereas Arab land holders attempted to qualify the lands as Miri, and so claim they had acquired it to prescription according to section 78 of the Ottoman Land law. (Kedar 1996, 300)

As discussed previously, under the Ottoman Land regime, the human voice was used as a unit of measurement between Miri and Mewat lands, so the new Israeli administration could easily overrule the outcome of such practices as imprecise and in need of more rigorous and modern approaches. Defining land as Mewat was thus in the interest of the state, and when disputes rose to the level of the

Supreme Court, the court also expanded the classification of what constituted Mewat land and in doing so classified land as state land in the majority of cases.

Various British and Israeli emergency laws—which, for instance, created closed military zones as well as changes to or non-enforcement of certain Ottoman and Mandate laws—further enabled the process of land transfer. For instance, the Israeli state changed the Ottoman law that Mewat land could be registered as Miri land after 10 years of cultivation. In order for Arab owners to claim ownership over land they now had to prove 15 to 20 years of cultivation. In 1961, the Supreme Court instigated an additional rule by which landholders had to have cultivated at least 50% of the land or it would become state land. Yet, how was one to prove whether peasants had cultivated more than 50% of their land? After 1961, the Supreme Court started to use aerial photographs (that were taken in 1945 by surveyors from the Survey of Palestine meant for mapping and developing Arab villages) as evidence to determine whether Arab landowners had cultivated more than 50% of their land. Aerial photographs hereby became another means by which Arab landowners could be dispossessed (Kedar 2001).

More generally, according to Sandy Kedar, it was "'a cultural clash of paradigms' in which the 'modern Western' legal system does not recognize the ways locals organize their spatial relations to land as giving rise to property rights" (2001, 927) that become a tool of dispossession. In land disputes, the Israeli Supreme Court did not accept the Ottoman practice of using the voice as a unit of measurement in order to define Miri land as this seemed far from being a clear and exact legal definition (Kedar 2001). Up-to-date techniques and methods for measurements, and concurrent expectations of exactness, therefore, became the means by which to challenge customary land allocation practices. The use of such allegedly more objective and scientific rules that did not account for the topographical and sociohistorical context helped bring about the appropriation of Arab land and its transformation into modern Israeli state land, that was surveyed, mapped, and parceled in terms of the new dominant ethno-national regime. Land transfer and expropriation hereby seemingly became methodological and technical, rather than political, issues (Kedar 2001).

The question of how much of the land that was transformed to state land was Arab land is deeply contested, with estimates varying between 4.2–5.5 million to 5.7–6.6 million dunams (Kark 1995). Given the blurry boundaries between different types of land and the various measurement methods used, ranging from the human voice to aerial photography and field surveys, such issues will inevitably remain deeply politicized. In any case, the facts on the ground were established and by the end of the land reform 93% of the land in Israel was managed by the Israel Land Administration (Kedar 1996).

Furthermore, the Israelization of lands was also the consequence of "national/collectivist elements in the Zionist vision" (Kedar 1996, 6). The socialist ideology that underpinned the Jewish project in Palestine and was embodied in the Kibbutzim (a communal, often agricultural, settlement) and the Yishuv served to enhance nationalist goals as well as the collectivist ethos. Socialist precepts not only provided legitimacy for transforming private land into state land but also became a means to encourage nationalist and collectivist sentiments. Besides redeeming the land for the socialist collective, the religious assumption that the Jewish people were called upon by God to redeem the Holy Land could also underlie the effort to do so. This combination of socialist, nationalistic, and religious ideals served to enhance Zionism's hegemony, which was so important to the Israeli state-building project. Ultimately, such hegemonic socialist-nationalist and religious sentiments enhanced the establishment of an ethno-national and ethnocratic state for the Jewish people (Yiftachel and Kedar 2000; Yiftachel 2006).

The transformation in the land regime in Israel/Palestine attests to how new legal precepts in tandem with science and technology helped establish a modern, territorially defined state. The law was employed to institute land reforms and to "Israelize" the land. Surveying and mapping with the latest measuring technologies were used to supersede indigenous, seemingly inexact and unscientific, land allocation practices. At the same time, Zionists' commitment to the modernist state-enforced ethos of order promoted their colonization project, both during the British Mandate and after the establishment of the Israeli state, to the detriment of the indigenous populations and their more traditional, informal and seemingly imprecise land management practices. The desire to impose top-down and state-centric order thus became an effective means to re-allocate land rights. Indeed, the Supreme Court's rejection of the voice as a measurement device to differentiate between Miri and Mewat lands

> can at least partially be reasonably attributed to its tendency towards formalism, to its desire to impose Western order on the perceived chaotic state of Ottoman land laws and to its preference for "objective" and clear-cut rules over blurry standards. (Kedar 1996, 303)

The Supreme Court thus strove for stability, order, and standardization, which went hand in hand with the Zionist project of redeeming the land (Kedar 1996). Indeed, the introduction of a legal system that gives little credence to its predecessors has historically been an effective tool of settler societies to dispossess indigenous populations of land rights (Kedar 2001; see also Porter, n.d.). While the Western scientific and legal paradigm enhanced the transfer of land, it also legitimized and depoliticized the new land regime and made it seem part of the natural order of things and an inevitable outcome of modernity.

4
Map-Making and Nation-Building
Pre- and Early State Mapping in Israel

Introduction: Maps at Work

Maps have many functions, yet in the process of building a nation-state one of their most important functions is to brand a territory to help create the imagined community of a nation. It is not only the shape of a bounded nation that helps create a sense of territorial belonging and national identity but also the textual and visual signifiers included in maps, which load them with national content. Nation-states have long used the tools of selective cartography to make a land their own. They may re-name the land and highlight certain features and sites that superimpose dominant national histories and narratives onto a land, while excluding others. When issued by the state, maps become a text that represents not a landscape, but a particular "hegemonic master narrative" (Elliott Weiss 2010, 203). How such narratives are distributed and targeted at particular audiences can influence their persuasiveness. Maps that become institutionalized and are circulated through national and transnational networks are more likely to have the last word in a war over territories. In this chapter we will focus on maps produced by Jewish organizations before 1948 and by Israeli institutions from the establishment of the Israeli state until 1967. The maps' spatial demarcations as well as the visual and textual signifiers that are included—and excluded—speak to the project of "judaizing" the territory and building a state.

For Baruch Kimmerling, Israel

> is a society plagued by the problem of existential legitimacy. It repeatedly has to explain to itself and to the international community why it chose Palestine, the land entitled "The Land of Israel," as its target territory for settlement. (Elliott Weiss 2010, 203)

Maps and their allusions to a biblical past, and the inclusion of a Hebrew topography, symbols, and names have therefore become part of the project of implanting Jewish culture into the soil of a territory.

The Politics of Maps. Christine Leuenberger and Izhak Schnell, Oxford University Press (2020).

DOI: 10.1093/oso/9780190076238.001.0001

The Jewish National Fund: Forging an Image of the Jewish Homeland

The Jewish Agency, in particular, was recognized by the League of Nations "as representing the interests of the Jewish diaspora in building the Jewish national home in Palestine" (Horowitz 1989). It, more than any other, in pre-1948 Palestine, become a transnational government in the making. The Jewish Agency provided resources for growth, demographic reinforcement, and the ideological justification for land acquisition, settlements, and national independence. Its main branch, the Jewish National Fund (JNF) (in Hebrew Keren Kayemeth LeIsrael [KKL]) exemplified the most official Jewish mapping practices in the pre-state era, which helped define the national territory.

Already in 1884 and then again at the first Zionist congress in 1897, Zionists suggested starting a national fund with the aim of purchasing lands in Eretz Israel (denoting an expansive biblical conception of the Land of Israel). The JNF, however, was not established until the Fifth Zionist Congress in Basel, Switzerland, in 1901. In order to fund the congress Haim Kleinman put together what he called an "Eretz Israel box" (also known as "the Blue Box"), as depicted in Figure 4.1.

The box was used to collect donations to buy land in the region and to increase public support and commitment to the Zionist project of building a Jewish homeland in the Holy Land. It became a symbol of the JNF. Blue boxes were

Figure 4.1. A sample of Blue Boxes. Available at https://www.kedem-auctions.com/. Accessed April 28, 2020..

distributed in communities, schools, synagogues, businesses, and homes all over the diaspora in North Africa, America, and Europe. They became an "expression of the deep bond between the diaspora Jewry and the small Jewish community and the soil of Eretz Israel." Indeed, as JNF describes it, "JNF made it possible for every Jew—whether man, woman or child—to become a partner in the Zionist enterprise and be personally involved in the development of the land."[1] Thus the Blue Box signified more than a need for donations, which could be made more efficiently by other means; it served to increase the diaspora's emotional involvement and support for the Zionist project and became one of the main agents of Zionist socialization.

The Blue Box become an icon, not only because it was placed in living rooms and community centers alongside other cherished Jewish symbols such as the Shabbat candles, but because it became part of rituals and ceremonies inside communities, schools, and families. It produced around itself a JNF culture (Bar-Gal 2003b) that had its own behavioral norms and ritualistic gatherings. It became the cornerstone at school festivals and fund-raising events. Indeed, the Blue Box became part of a civil religion that was an integral part of the diaspora community. It helped constitute the collective memory of being a Zionist and gave meaning to the concept of being Israeli.

Until the arrival of the Blue Box in living rooms, classrooms, and community centers all over the diaspora, the concept of the Land of Israel was esoteric and abstract to most. Some may have dreamed of a messianic miracle that would bring about redemption of the Holy Land. But mostly the Jewish collective identity was not tied to a territory but rather was forged by a strong sense of collective identity and a dedication to commonly shared traditions. Even the founder of the Zionist state-building project, Theodor Herzl, in his 1895 book *A Jewish State*, had such little attachment to the mystical image of the Land of Israel that he urged the Zionist movement to consider Uganda as a home for the European Jewry. The intermarriage of Jewish collective identity with a territory was thus slowly forged, with the Blue Box at its core. While early versions of the Blue Box depicted only the Jewish symbol of the Star of David, later versions had it side by side with a map of Eretz Israel so as to fuse the association of Jewish identity with a territory. Only after the 1930s did the Star of David disappear from the boxes to be replaced only by a map. Indeed, at that point, Jewishness had seemingly become firmly tied to a territory. Both the Star of David and Eretz Israel could now independently and forcefully serve as widely recognized signifiers of Jewishness. JNF's attempt to brand the territory as the Jewish homeland had seemingly succeeded.

[1] Jewish National Fund, available at www.jnf.org. Retrieved April 28, 2020.

The JNF, alongside many other institutions and organizations, followed the trends at the time of increasingly using maps for propaganda purposes. Maps therefore also seemed to provide the answer to JNF's challenges, such as how to justify its own attempt to purchase and nationalize land in the Holy Land and how to convey the boundaries of the Land of Israel and market their product: Eretz Israel. Such rhetorical means were particularly important given that JNF had to justify its approach to American Zionists, who had established an additional fund, Keren Hayesod. Its protagonists, in line with American ideas of free enterprise and individualized land ownership, did not believe in purchasing and nationalizing land, but they believed in supporting private settlements of Jews in Palestine based on individual land purchases. JNF's maps therefore tended to visually emphasize its seemingly large landholdings as opposed to the more scattered patches of land acquired by Keren Hayesod, which would design maps highlighting their settlement activities in turn. The JNF, however, was not only battling the American Zionists over strategies about how to redeem Eretz Israel but also had to sustain its ideological visions in an international arena in which political powers were frequently re-naming, re-defining, and re-drawing boundaries and territories east of the Mediterranean Sea and deep into present-day Jordan.

At the end of World War I, the head of the Zionist Movement, Chaim Weizman, expressed his wish that the Jewish homeland "should occupy both sides of the Jordan, and should include (in addition to the land west of the Jordan) areas that are today a part of South Lebanon (Tsiddan), the Golan Heights and the Kingdom of Jordan" (Bar-Gal 2003b, 4). It is that territorial vision that was to be imprinted on Blue Boxes and is represented on JNF's propaganda material to this day, as seen in Figure 4.2.

Maps of the Jewish homeland were thus cartographically anchored by the length of the Jordan River, that was centrally located with two-thirds of the land mass to the west and one-third of the territorial land mass to the east of the river, with only the Mediterranean Sea to the west and the Blue Box's rim to the east, bounding what was considered to be the "Promised Land." Two main vertical lines typify the territorial image: the softly curved Mediterranean shoreline and the Jordan River including the Sea of Galilee, also known as Lake Kinneret, and the Dead Sea. The inclusion of the Jordan River and Jerusalem speak to the map/s allusion to biblical prophecies. The Jordan was to symbolize the entrance gate of the people of Israel to their biblical land, and Jerusalem always served as a powerful symbol of the messianic longing for the Holy Land. The very scarcity of lines and information served the map's function to be as accessible, uncluttered, and visually powerful as possible so as to imprint itself like a logo on its target audience that was to identify with the land and its Jewish history and future.

Figure 4.2. The "Blue Box" produced by the Jewish National Fund from 1930 onward.

One of the defining characteristics of JNF-produced maps was their largely "Zionist-focused cartographic selection" (Bar-Gal 2003b, 13). The assumption was that the maps' target audience were Jews who either needed to be educated in the land's Jewish heritage and significance or visually persuaded to provide donations and public support to JNF's mission. Therefore, Hebrew settlements, sites, and historical places were predominantly marked, while Arab places, sites, and settlements were largely excluded. As a result, the superimposition of Hebrew and Zionist symbols on a white area that was seemingly empty and barren, with no settlements or inhabitants, gave the impression that this indeed was a virgin land ripe for fertilization and colonization.

In 1922, however, the British Mandate government used the length of the Jordan River, with the Lake Kinneret to the north and the Dead Sea to the south, as the boundary between the two territories; the territory to the west was, according to the 1917 Balfour Declaration, to became the Jewish national home in Palestine and, to the east was Transjordan.[2] In its partition plan in 1947 the

[2] See Israeli Ministry of Foreign Affairs's website www.mfa.gov.il for the 1922 map of the British partition of the land.

United Nations divided the territory accordingly and assigned much of the West Bank to an unnamed Palestinian "Arab State" (Galnoor 1995, 284). The willingness of the then politically dominant Labor Zionists to accept the UN's partition plan was largely pragmatic. Its acceptance capitalized on the goodwill of the international community and provided the Jewish community with a territory they could call their own, yet the agreement was not set in stone. As Ben Gurion pointed out, "History makes no final arrangements," and regimes and borders can always be changed (Tarrow 2013, 157).

However, this territorial compromise was contested by both Revisionist Zionists, who claimed everything from present-day Israel, the West Bank, Gaza, Jordan, and southern Lebanon, to religious Zionists who, following the Bible, laid claim on land extending from Egypt to the river Euphrates, including present-day Israel, the West Bank, Gaza Strip, Jordan, Lebanon, and parts of Egypt, Syria, Iraq, and Turkey. Given the diversity of opinions on matters of territories—ranging from the Labor Zionists to the revisionists and the religious Zionists, the JNF chose to represent the maximalist territorial image of the homeland avoiding demarking any boundaries. Even though the Zionist movement gave up the territory east of the Jordan River and agreed to a smaller and more limited land mass for the Jewish people, the JNF refused to amputate Eastern Trans-Jordan from the area which was supposed to be part of the Jewish Homeland (Bar-Gal 2003a, 1999, 2003b). Thus despite these internal political struggles, the JNF continued to represent a Zionist consensus with which Jews the world over could identify: the map on the Blue Box kept representing Greater Israel with its boundaries undetermined.

Even the declaration of the state of Israel in 1948, which had no permanent boundaries (but was framed by temporary ceasefire lines), did not impact the JNF's cartographic norms for branding the Jewish homeland. Despite changing territorial circumstances, boundaries, and the appearance and disappearance of ceasefire lines, the logo map of the Jewish homeland remained the same. Sixty years later, a JNF poster, as shown in Figure 4.3, represented an early Blue Box arising out of JNF-planted forests.

Indeed upon the state's sixtieth anniversary, JNF was celebrating its own successes in collecting donations, purchasing and nationalizing land, and greening a country. Neither international agreements, the delineation of international borders, nor changing realities on the ground prompted JNF to change its map of the Jewish homeland. Its brand name and territorial delineation remained the same, providing Jews in the diaspora a stable and fixed logo of what, according to the JNF, is and always should be the shape of the Jewish homeland.

While the Blue Box was one of the more powerful Zionist icons, the JNF also produced a range of other products to propagate the Zionist vision and its

Figure 4.3. Jewish National Fund's 60th Anniversary Poster of the State of Israel.
Source: Jewish National Fund.

historical narrative. These include stamps containing miniature maps, images of Zionist cultural and national symbols, books, games, and films. Educational games and stamps in particular used maps to forge in its target audience a sense of the unbounded territory of Eretz Israel and spreading the ideas of Zionism and the settlement of the land.[3] For instance, the JNF produced games (as shown in Figures 4.4a and 4.4b), in which players threw dice to move their pieces along the Jewish colonies, buying land along the way. Upon completing a successful purchase they would get an ownership card that detailed the story of the colony purchased.

Arieh El-Hanani (Sapozhnikov) produced a revisionist map of the territory. His map was also inspired by Chaim Weizman's demand for a Jewish homeland that should be extensive enough to control its own water resources (as shown in Figure 4.5). His revisionist stance on territorial issues appealed to the JNF's leadership, which reproduced the map on stamps.

The map included such water sources as the Jordan and the Litani rivers (in south Lebanon) within the Jewish homeland. The map "judaizes" (Zink 2009, 122) the land, as most Arab sites and settlements were excluded (except Gaza, Hebron, and Shchem [Nablus]). Instead, Zionist symbols decorate the landscape, such as a farmer plowing the fields, and giant orange and olive trees; also Jewish archaeological sites are foregrounded such as David's Tower, Rachel's Tomb, and the Patriach's Tomb; mountains and regions were given biblical and Hebrew names so as to infuse the topography and the land with a historic value to Jewish civilization. This map represents a vision of a Greater Israel that some present-day settlers' rabbis continue to advocate.

The JNF aimed not only to socialize a people into a notion of a territory and its imagined community through games and stamps, but it also became increasingly involved in the formal Hebrew educational system. Besides such propaganda maps, JNF also produced seemingly more scientific maps. By 1928, the JNF designated the cartographer Abraham Y. Brawer to design a map for use in schools. Brawer was to become one of the founders of the discipline of geography in Israel and he was also known for his maximalist territorial vision. Brawer adopted Maimonides's (also known as Rabbi Mosheh ben Maimon, or, for short, Rambam) interpretation of the biblical boundaries of the Land of Israel, comprising the territory in which the 12 Jewish tribes (ranging from Asher to Gad) had settled and in which the commandments to redeem the land could be fulfilled. Thus Brawer's map included both sides of the Jordan River as it was there where Joshua allegedly commanded the tribes to settle (Figure 4.6).

According to this map, the Jewish homeland does not include the coastal plains north of Haifa and south of Ashdod as they were populated by the ancient Siddons and Philistines. This map also does not mark any borders, and so

[3] For further information see World Zionist Organization: The Central Zionist Archives, Childrens Games, http://www.zionistarchives.org.il/en/AttheCZA/Pages/Games.aspx. Retrieved May 28, 2019.

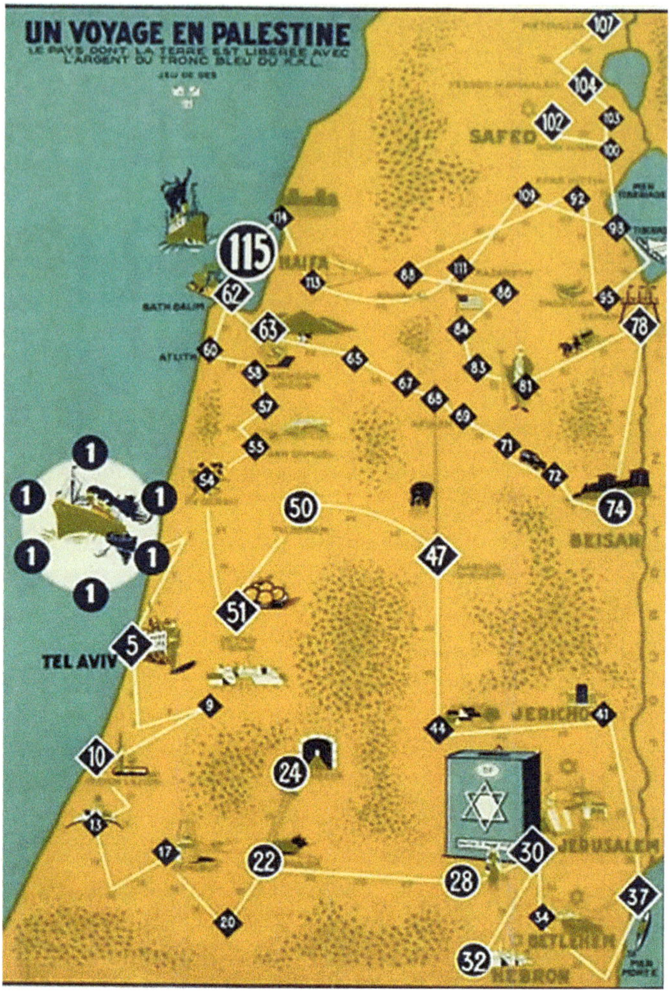

Figure 4.4a. Jewish National Fund board game: Un Voyage en Palestine: Le pays dont la terre est liberee avec l'argent du tronc blue du K.K.L. ([Take] A voyage in Palestine: A country whose land was liberated with money from the Blue Box of the Jewish National Fund), ca. 1950. Available at https://www.palestineposterproject.org/poster/un-voyage-en-palestine. Retrieved April 28, 2020.

leaves the extent of the Jewish homeland undetermined. Brawer's map exemplified an era. Indeed, before the British Mandate was established, it was standard practice to cartographically representing Eretz Israel within its Biblical borders (Bar-Gal 2002).

Figure 4.4b. A Trip across the Country, 1946 Board Game. Available at https://www.palestineposterproject.org/poster/journey-through-the-land. Retrieved May 28, 2019.

Yosef Azaryahu (1929), the supervisor of education in the department of education who formulated the geography curriculum for the Hebrew educational system, assessed JNF's maps for their use in Hebrew schools. He demanded that Jewish land holdings (whether they were JNF's land holdings or privately owned Jewish properties) be demarcated in strong colors to clearly distinguish them from the surrounding topographical features. He also stipulated that the

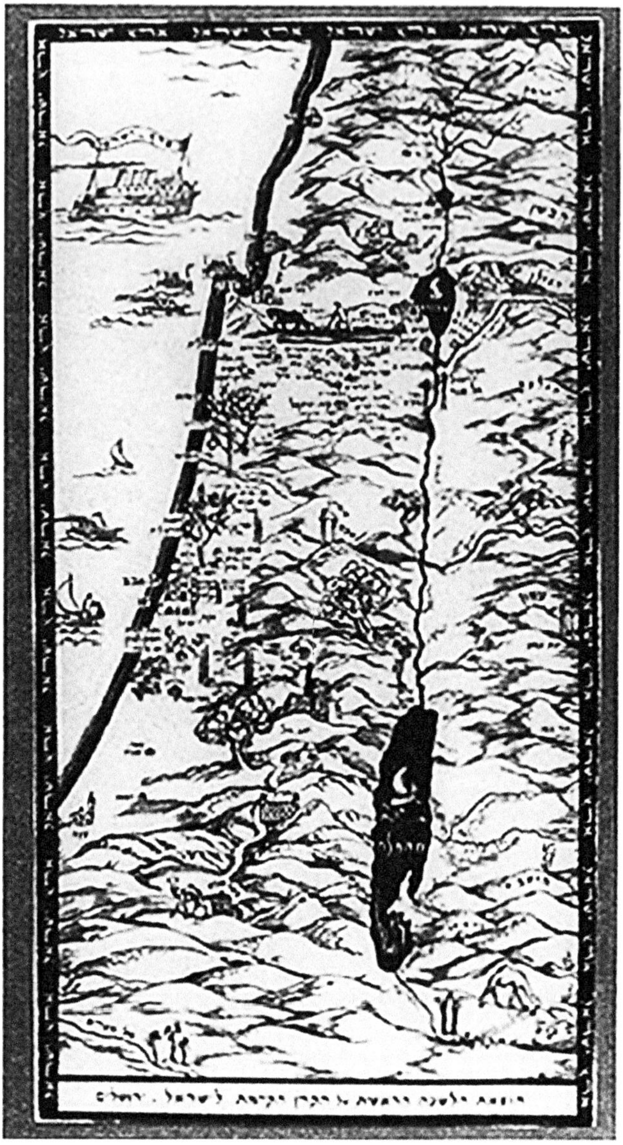

Figure 4.5. Arieh El-Hanani (Sapozhnikov) map from 1930 on a Jewish National Fund stamp. Available at https://www.palestineposterproject.org/poster/pictorial-stamp-map. Retrieved May 28, 2019.

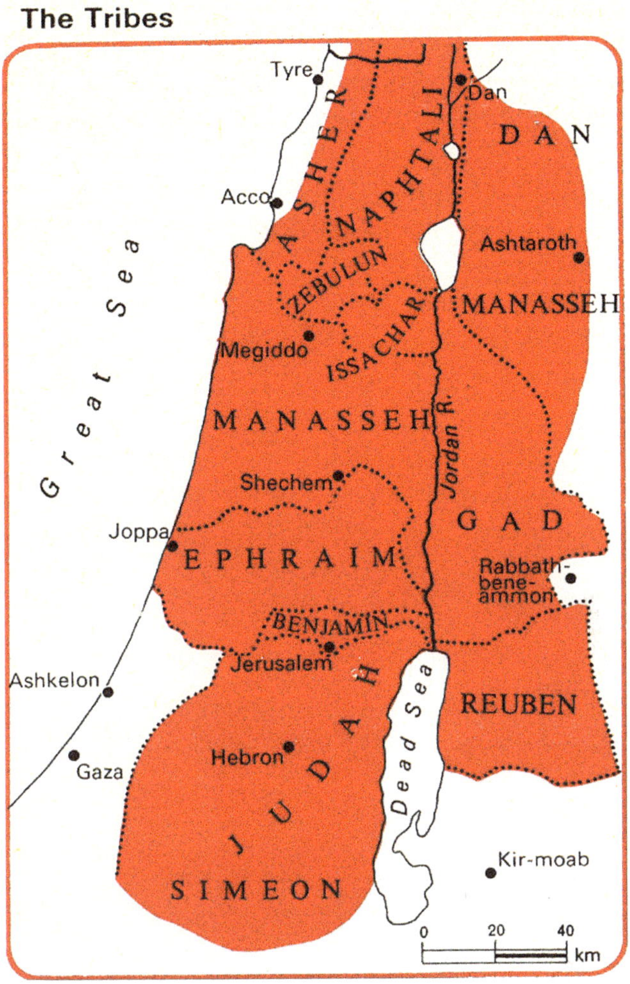

Figure 4.6. "The Tribes," in Carta (1977), *Carta's Historical Atlas of Israel*. Jerusalem: Carta, 8.

maps include the current political boundaries. JNF revised their more scientific maps accordingly yet continued to differentiate JNF land holdings from privately owned Jewish landholdings on propaganda maps so as to emphasize its own role in redeeming the Land of Israel. By presenting a maximalist territorial interpretation of Eretz Israel, JNF continued to present an alternative territorial vision to the bordered territory of Mandatory Palestine. With such maps, JNF shaped "the ideological template of the Israeli education system" (Weiss 2010, 206).

The ideology and politics that permeated such maps is exemplified by a JNF leaflet that was produced on the eve of the establishment of the Israeli state in 1947. The leaflet contained three topographical maps that summarized JNF's progress in terms of purchasing and nationalizing landholdings in Eretz Israel (Figure 4.7).

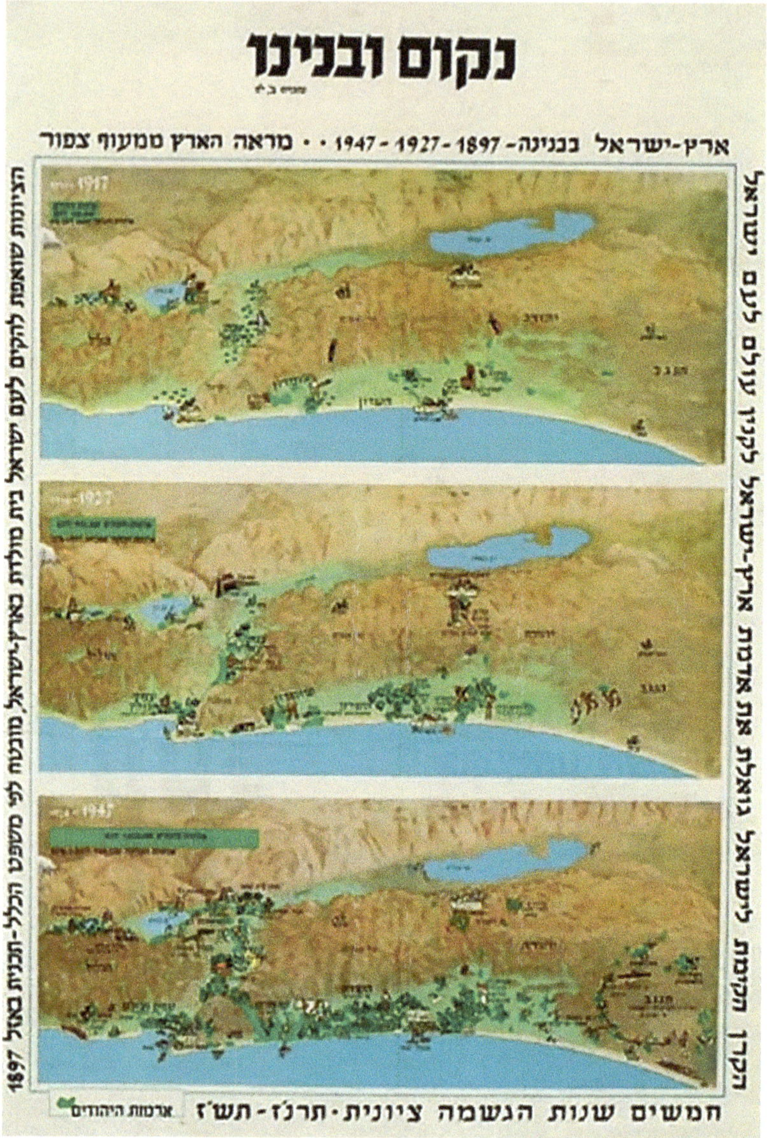

Figure 4.7. Jewish National Fund, 1947, The Eternal Property of the Jewish People. Available at https://www.palestineposterproject.org/poster/the-eternal-property-of-the-jewish-people. Retrieved June 5, 2019.

The three maps visually represent the increase in Jewish settlements and the progressive greening of the land of Israel as it came under Jewish ownership. The cartographic emphasis on the greening of the country signified the fertility of the land and celebrated the Zionist settlers' achievements in its cultivation. Similar to other maps, Arab sites, places, and towns were excluded, which left the impression that the growing green spaces were expanding on a formerly abandoned, barren, and uncultivated frontier. The names given to the regions are biblical, which connected JNF's redemption of the land with the religious command for the Jewish people to redeem the land for the Israelites. These visual clues are juxtaposed to its textual framing. The leaflet's heading contains the Prophet Nechemia (2,18) statement "Arise and let us build," while the maps are framed by statements such as "The KKL (JNF) is redeeming the lands of Eretz Israel as an eternal property of the people of Israel"; "The Zionist movement seeks to establish for the people of Israel a homeland in Eretz Israel as it was promised to the public in the Basel plan—1897"; "Eretz Israel in its building 1897—1927—1947—The country from a bird view"; and "Fifty years of Zionist accomplishment 1897 to 1947." The text in the box states "Jewish lands" (referring to green areas on the map). These maps and their textual framing exemplify JNF's ideology and mission and its ability to impart in its audience the sense that its aims were not only those of the Jewish nation but also the fulfillment of biblical commandments.

As we discussed in chapter 2, the Jewish community in Mandatory Palestine constituted a state-in-the-making. Institutions, such as the JNF, became a crucial part of the state-making process. Their distribution of maps of the Jewish homeland on Blue Boxes, stamps, posters, board games, and educational material served as branding logos that were to foster the link between a territory and a people. Through maps, rituals, and practices, the land came to signify the essence of Jewishness—it, like the Shabbat candles and the Blue Box, became the core of what it meant to be a Zionist. The territory also was "sacred" as visions of Eretz Israel became intimately linked with biblical prophecies. Indeed, it was now that the Jewish people had to follow Nechemia prophecies to arise and build on the land that seemingly had been abandoned, yet had been awaiting Jewish redemption for 2,000 years. And it was the map on the Blue Box, depicting the lands bought, settled, and cultivated that made every man, woman, or child who donated money a partner in the common project of Jewish redemption in the Holy Land. Just as in Siam, Rhodesia, and Burma where maps crucially consolidated national identity (Anderson 1983; Winichakul 1994; Collier 2006), JNF's logo maps on Blue Boxes, games, and posters helped transform Jewish identity into a national identity rooted in a territory that became associated with the Jewish Golden Age.

Early Statehood Maps: What's in a Name?

National movements have historically always used names to make a land their own. After gaining control of a territory, surveyors assert legal ownership through measuring and mapping out the field while governmental naming commissions stake out the territory by naming sites, landscapes, mountains, and towns. As David Day points (2008) out,

> Just as naming a place and its features is one of the first acts that people do when taking possession of an empty land, so is renaming those features one of the first acts that supplanting societies do when they embark upon the prolonged process of claiming the territory of another society as their own. (Day 2008, 50)

Re-labeling and re-naming is a pre-condition for the transfer of territorial control and its management. Therefore, supplanting societies re-name newly acquired territories each time anew. It is a cultural project that aims to fulfill the colonizers' mission of "internal colonialism" (J. C. Scott, Tehranian, and Mathias 2002, 26) by establishing an ethno-national typography of the land that makes the landscape newly legible.

A state's project of making a land legible frequently overlies, and at times, supersedes and replaces the often fundamentally different, local, and customary naming practices. While local naming practices encode important local knowledge (such as that "Durham road" leads to Durham) and suit the needs of knowledgeable locals within a particular community, state naming practices require "a synoptic view, a standardized scheme of identification" (J. C. Scott, Tehranian, and Mathias 2002, 5). Through the use of standardized administrative techniques, a state's proxies construct a guide for official strangers, planners, tax collectors, transportation managers, and police officers from different localities. For the purposes of the state, the inconsistency, messiness, and multi-layeredness of local naming conventions are nothing but an obstacle to effective rule.

Consequently, naming commissions tend to produce a national toponymic system that reflects a set of national and cultural values, meanings, and histories that can help unify a territory and cement a sense of national belonging (M. Azaryahu and Kook 2002; Caplan and Torpey 2001; Cohen and Kliot 1992; Jackson 1970; Kadmon 2004; Lewis 1982; Rose-Redwood, Alderman, and Azaryahu 2010; Zelinsky 1984, 1986; M. Azaryahu 1996). Re-naming a territory using a taxonomy that represents the national culture is particularly important for states in the making in search of legitimacy (Zelinsky 1984, 1986). In struggles over territories, map wars historically have gone hand in hand with name wars. Adopting new place names and rejecting old ones is a characteristic feature of conflicts over contested lands. After radical regime change, streets and squares are frequently re-named to instill the symbolic power of the new regime and displace the old. In the 19th century, the systematic construction of national toponomies became an integral part of displacing an indigenous people and building the structure of a new nation-state in its place.

After the establishment of the Israeli state in 1948, the government brought about an onomastic revolution. Everything from family names to place names was to be remade so as to culturally engineer a new Hebrew culture. As part of this effort, the Hebrew language was revived. This common language was to be a lever for culturally unifying a people from different diasporas, ranging from Africa to Europe, in order to create a Hebrew civilization with its history rooted in the biblical homeland (Schweid 1995). As part of this project, Israel's first prime minister David Ben Gurion also instituted the Hebraizing of people's names. People with names from such diverse places as Germany or North Africa were encouraged to adopt new Hebrew names from Ze'ev to Ari and from Ben-David to Abramowitz. For candidates for official positions, such as high-ranking army officers and diplomats, Hebraizing their name was a pre-condition for service.

Re-naming the landscape also became a major part of the Zionist culture and nation-building project. The struggle to Hebraize the map had already started in 1925, when a special committee was established to assign Hebrew names to new Jewish settlements. These names were frequently inspired by Hebrew, biblical, Talmudic, or Zionist historical and nationalist sources. National movements have always used historic names, particularly the names of national symbols and heroes, to remake a toponomy (Cohen and Kliot 1981; Derzia 1993; Bitan 1992). Not surprisingly, then, early naming practices in Israel also consisted of deliberate efforts to imbue the map with names to foster people's identification and attachment to the land by invoking biblical history and calling upon Zionist nationalist symbols. For instance, names borrowed from the Old Testament continue to mark the geography of Israel today: "King David Street" in Jerusalem and the hills of "Judea and Samaria" in the West Bank invoke early Jewish presence in the region.

With the establishment of the Israeli state in 1948, Prime Minister David Ben-Gurion appointed a Governmental Names Committee to establish a Hebrew typonomy of the land (Srebro 2009; El-Haj 2002; Bier 2017). Re-making the map was seen as fundamental to forging the new nation. According to the interior minister at the time, in order to have a "people that is rooted in the soil of our country, we must begin in the fundamental Hebraicization of our country's map" (M. Azaryahu and Golan 2001, 184). At the same time, to "redeem" Israeli land "from the rule of foreign language" was also a means to establish Jewish sovereignty over the territory (M. Azaryahu and Golan 2001, 187). Indeed, in a letter to the committee, Ben Gurion pointed to the importance of re-naming the Negev region,

> You have banished the shame of foreignness and of an alien language from half of Israeli territory and completed the job begun by the Israeli defense forces: to liberate the Negev from foreign rule. (M. Azaryahu and Golan 2001, 187)

During the 1940s, the Survey of Palestine had provided an authoritative 1:100,000 map (according to which 1cm on the map equals 1 km on the ground) with more than 3,700 Arabic names and just over 200 Hebrew names for Jewish settlements, as well as some Christian names. These Arab and Christian names, however, were considered foreign from the perspective of the Zionist community. Their assumption was that although only 174 toponymical items were mentioned in the Old Testament and other historical sources, there existed a Hebrew toponymy. Thus, many Arabic names were thought to be a modification of original Semitic names. It was the Arabization of the land and its names that needed to be reversed by instead using new names to invoke a Hebraized geography.

Mountains, valleys, springs, rivers, roads, and settlements were re-named in order to establish a Hebrew topography of the land. For David Benvenisti, the producer of the national map, Hebraizing the map was a sacred task. As Benvenisti's son attests:

> Changing place names in order to arrive at a Hebrew map of Eretz Israel was considered by my father a sacred task. Renaming the landscape was a way to reestablish contact with those same landscapes and places from which we had been physically removed for 2000 years but whose names we had always preserved. We carried with us for centuries our *geographia sacra*, not only Biblical names but all the Mishnaic and Talmudic names. (Benvenisti 1989, 11)

In line with such national and cultural sentiments, the Governmental Names Committee instituted several principles. Most important, providing Hebrew place names was paramount. Non-Hebrew names, also those of notable Jewish Zionists, were dismissed, so as to maintain a purely Hebraized toponymic map; biblical or Talmudic place names had the highest priority. As the Bible mentioned relatively few place names, however, they could not be used for more than 20% of the names needed. Instead, new archaeological sites that sprang up as a result of Bible-based archaeological digs could at times supply more biblical or Talmudic names. Moreover, the committee at times used Arab names when they possessed a similar phonetic structure to an ancient Hebrew name (as is the case for the Arab town of Sachnin which is similar to the ancient Hebrew place name

Sichnin). The assumption was that their similarity resulted from the continuity in the name's use during the Jewish, Byzantine, and Muslim eras. Furthermore, the committee preferred names that connected to Jewish history or celebrated the Zionist ethos. Arab towns and villages, however, were the exception to the rule. Arab site names were to be eliminated and Arab villages and towns that were destroyed during the 1948 war were to be cartographically erased. Yet, the little more than 100 still populated Arab villages and towns in the country's periphery maintained their Arab names, signifying their foreignness. Their limited number, however, could not change the Hebraized character of the new toponymic map of Israel. At the same time, the presence of Bedouin settlements and land use was neither legally nor cartographically recognized, with disputes over their land rights continuing to this day (Yiftachel 2006). Thereby the national map marked the beginning of the "Hebraization" and "Israelization" of the land, and the "designification" and symbolic erasure of Arab and Bedouin spaces (Collins-Kreiner, Mansfeld, and Kliot 2006; Falah 1989, 1996; Benvenisti 1989; Newman 2001).

The same general guidelines that the first Governmental Names Committee had instituted during Ben Gurion's administration continued to inform the committee's naming practices for more than 60 years. Hanna Bitan (1998, 1992), the committee's longtime secretary, emphasized that the committee members never implemented any formal decision about the exact criteria to be used in choosing and assigning names. No official document containing formal guidelines was ever produced or ratified. Over the years, however, an unwritten consensus emerged that biblical, Talmudic, and Zionist names would be prioritized over names based on nature or Hebraized local names. The new revolutionary toponymic map was thus not a product of formal regulation or instruction, but it was implemented by experts who had developed a consensus over what constitutes best practice based on informal guidelines, interactional contingencies, and institutional cultures (Leuenberger 2012). Arguably, these experts, like the early pioneers in the Yishuv, were "recruited" to the national ethos and a vision of a new society and a new geography that was to further the Zionist nation-building project. By the end of the 1950's, the committee published two maps at the scales of 1:100,000 and 1:250,000 (according to which 1 cm on the map corresponds to 2.5 km on the ground) with primarily Hebrew names. By the 1950s there were just around 3,000 Hebraized names; by the 1960s this number had increased to 5,000, and by 2000, over 9,000 sites, settlements, landscapes, and streets had been re-named (Cohen and Kliot 1992; M. Azaryahu 1996; M. Azaryahu and Golan 2001; M. Azaryahu and Kook 2002).

Upon the Hebraization of the national map, the government engaged in deliberate efforts to disseminate the new Hebrew toponymic map so as to assure the adoption of the new names in everyday use. All new names were registered and used by the Survey of Israel. David Ben Gurion's administration also instructed the military to use only Hebrew place names so as to counter soldiers' tendencies to use customary Arab place names. On the ground, too, road signs were replaced with new Hebrew names inscribed on them. Likewise, schools adopted the updated maps and atlases; the Hebraized maps were to "influence the schools, their teachers and pupils, to take upon themselves the task to uproot the foreign names and to root the Hebrew names" (M. Azaryahu and Golan 2001, 188). For instance, after the Ministry of Education approved the new Hebrew map entitled "Eretz Israel: A Map for Knowing the Homeland for Youngsters and the General Public,"[4] this map was to be placed in elementary schools throughout the country (as seen in Figure 4.8) and it hung on classroom walls up until the 1967 war (Berger 2008). The map's name indicates that it defines not a nation-state but the territorial base of the Jewish homeland. With an immigrant population from all over the diaspora ranging from the Arab world to Germany, defining the homeland was a crucial aspect of re-socializing the population into a shared national history.

This map exemplifies official mapping practices in Israel before 1967 in terms of how it demarcates boundaries, designifies Arab topography, and Hebraizes place names. The boundaries delineated here correspond to the colonial boundaries demarcated by Britain and France and ratified by the League of Nations when it approved the British Mandate over Palestine in 1922. However, while the internationally recognized boundaries are delineated (as a sequence of black plus signs), they visually recede into the background compared to the visually prominent red lines that indicate the armistice lines between Israel and its Arab neighbors agreed upon in 1949. Thus, where the international boundary diverges from the red ceasefire lines, the armistice line visually predominates along the Gaza Strip. While the international boundary markings with Transjordan are eliminated, an armistice line marks the boundary between Israel and the West Bank. The West Bank is presented as part of Jordan, erasing its unique status as a separate, yet unnamed Palestinian state with a Palestinian territorial base for a Palestinian people. In Gaza, too (even though it was never formally annexed by Egypt), the territories' Palestinian-Arab identity was cartographically erased.

The politics of erasure is underscored by excluding the UN's 1947 partition of Israel/Palestine (which provided international legitimacy to two states within historic Palestine) from the map. The elimination of the 1947 UN partition plan, the visual marginalization of the international boundaries, and the visual prominence of the ceasefire lines, function not only to signify that Israel's boundaries

[4] Translation by Izhak Schnell.

Figure 4.8. The Land of Israel: A Map for Knowing the Homeland for Youth and for the People (Ben-Tsiyon and Benvenisti, n.d.). Courtesy of Harvard Map Collection.

are provisional and temporary, but also to cartographically open up the West Bank as a potentially Jewish space and new frontier. Physical infrastructure such as roads that cross the ceasefire line only highlight the vulnerability of the West Bank's status as a distinct, separate, and Palestinian-Arab territory. Such cartographic representations left the question of the whereabouts of the borders between Israel, the West Bank, and Jordan unresolved and opened up these territorial spaces to Jewish dreams for lost frontiers. Indeed, geography lessons at the time included the study of the East and West Bank of the Jordan as regions of Israel; the West Bank was re-named with its biblical names, Judea and Samaria, and the study of their physical characteristics made them part of the Israeli territorial imagination.

This map also cartographically exemplifies the "de-Arabization" of the land. Arab towns and villages that were evacuated and destroyed by the Israeli army in 1948 were eliminated. In many of these sites, the Israeli government had settled Jewish refugees from the Holocaust and other Arab countries into new population centers that only rarely had names similar to their original Arab location and were Hebraized. Also, Bedouin living spaces and land use in the Negev and the Galilee were not included in the map. Their more temporary residential structures (such as huts and tents) were not recognized by the Israeli government as formal settlements, especially as the Bedouins were considered to be residing on what the government had defined as state land. Only those Arab villages and towns that remained populated after the establishment of the state were included in the map with their original Arab names. Consequently, the new map created an image of the homeland with a Jewish majority residing in Jewish towns, and with a small minority of Arabs living mostly along Israel's national periphery.

The cartographic erasure of Palestinian Arab spaces and places was part of a state-initiated attempt to expunge the pre-1948 non-Jewish history from the region. It was a process of forgetting and narrative reconstitution that marginalized or eliminated one set of histories over others. Such historical reconstructions were helped by the fact that post-1948 the Israeli military destroyed former Arab villages and towns, and Israeli school textbooks either erased from history the flood of over 800,000 Palestinian refugees, or presented the Palestinians' flight as a voluntary exodus (Schnell 2002). It was not until 1980 that exceptions to the rule began to appear. For instance, the *Carta Atlas for the History of the State of Israel* (Carta 1980) included one map appertaining to formerly inhabited but subsequently evacuated Arab towns and villages. In the 1980s, Israeli activists, Palestinian residents of Israel and the West Bank, and Palestinians in the diaspora also started to design maps and collect stories to highlight the history of the Palestinian people before 1948, including memories of Israel's war for independence, an event Palestinians remember as the Nakba.

Despite the fact that the Governmental Names Committee's naming preferences did not change substantially over the years, there were nevertheless differences in naming practices. For instance, the naming of streets became less of a top-down than a bottom-up process. As a result, the politics of street names differs from the politics of naming on the national map. Instead of the Governmental Names Committee being in charge, it is the local municipalities that have the power to name within their area. Top-down ideological directives thus mix with local politics in different towns. Political party alliances and the ethnic composition of a town (whether it is predominantly Jewish, Arab, or ethnically mixed) co-determine the naming practices at the local level. Indeed, the naming of streets can become a source of intense debate in municipal councils (Bar-Gal 1992). Various publics, with different political affiliations, ranging from the revisionists, Labor Zionists, the National Religious and the Ultra-Orthodox, can become part of debates over naming practices.

The ethnic composition of towns and cities can co-determine the sort of naming conventions implemented by different municipalities. For example, the Tel Aviv and Haifa municipalities frequently commemorate Zionist political and cultural leaders by their Hebrew names. Commonly used names are those of individuals who rose above sectarian divisions and gained national recognition, such as Thedor Herzl, the visionary of the Zionist movement; Chaim Weitzman, the first Israeli president; and Chaim Bialik, the Israeli national poet. However, local politics can also impact the decision to name streets after more contentious figures, such as the rivisionist Ze'ev Jabotinski. Also, in the ultra-orthodox city of Bnei Brak, the municipality instituted naming preferences based not on Zionist leaders but on ultra-orthodox figures, such as Rabbi Shach. Yet, in some places such as Jerusalem, Rehovot, and Netivot, local leaders tend to be commemorated by having their names as the bulk of street labels. Also, unlike the Governmental Names Committee, municipalities assign names to streets, squares, and other sites to commemorate leading figures associated with the diaspora whose names are meaningful to the dominant immigrant group in their locale.

The discrepancy between top-down and bottom-up naming becomes particularly stark in the case of Arab Israeli towns within Israel. Traditionally, the streets in these towns did not have names. The practice of naming streets was only introduced slowly and intermittently. Often streets were numbered instead of named. However, in several major Israeli-Arab townships, such as Nazareth and Um-el-Fachem, the municipality named some of the main streets (M. Azaryahu and Kook 2002). It is only in Um-el-Fachem that the municipality used names in order to resist the dominant Zionist ethos by composing a comprehensive toponymic map of street names. There, most street names commemorate either Islamic cultural and political leaders, major battles won by Muslims, or local military leaders such as Izz-Al-Din-Al-Kassam, but they exclude leaders from the

Palestinian national movement. The names chosen speak to a desire to construct a Muslim identity that goes back to Muslim Spain and the Islamic empire in its Golden Age. On the other hand, in Nazareth, which is a stronghold of Israel's communist party, a number of streets carry the names of prominent local communist party members. These naming practices exemplify the attempt of Israeli Arabs to ascertain a national and cultural identity within the predominant ethnocentric Jewish sociocultural system.

As we have discussed earlier, state naming practices and local customary naming practices have always been strikingly different. While each set of names is designed to make a landscape and townscape legible, state naming practices become part of standardized administrative techniques (J. C. Scott, Tehranian, and Mathias 2002) that serve the functioning of the state system. Local naming practices, on the other hand, embody valuable local knowledge that may resist top-down schemas of administrative rationalities. While the purpose of state-administered naming practices is to make a population and a territory legible within opaque and complex local realities there always remain spaces of illegibility. Indeed, as unnamed streets escape the logic of the state's rationalities they become a maze that can only be navigated with local knowledge as a compass. Israel's many street names reflect the often short arm of the Jewish Zionist state, and they also reveal the intermittent bottom-up resistance to its hegemonic vision. In most places, though, the long arm of the state has molded a territory and a people into a common polity with a common vision. Most municipalities followed the nationalistic project of Hebraizing the toponymic maps of their cities in line with the Zionist ethos and legacy.

It is worth noting that the main principles for naming places pre-1967 have hardly changed in post-1967 Israel. Indeed, the Governmental Names Committee employed some of the same naming conventions to Hebraize the West Bank, Gaza, and the Golan Heights so as to construct toponymical consistence over what they perceived to be the whole of the Land of Israel. The names assigned to places within the oPt were most often biblical; also more "natural" and "abstract" names tended to revive biblical places and symbolize the return to tribal homelands, such as Matzadot-Yehuda (Judea's Strongholds) and Mekhorah (Native-land). As the present-day West Bank territory was perceived to have been the core settlement area during the biblical period, such biblically inspired names were ascribed to about two-thirds of the settlements in Judea. On the other hand, Talmudic names were more frequently used in Galilee as Israelites settled this region during the Talmudic period, and there was an ample supply of such names. Over time, naming preferences included more and more natural and abstract names and were frequently used in areas inside the Green Line which demarcated the 1949 armistice lines. Moreover, the types of settlements could also determine naming practices. Settlements associated with the labor movement, for instance, rarely used biblical and Talmudic names, but they were likely to draw on nature or Zionism as sources of inspiration. The religious and political makeup of a settlement could thus co-determine predominant local naming practices. As we have seen, re-naming the territory is a national project that extends from the national map to street names and in this way becomes intertwined with national priorities and local politics.

Every State Needs an Atlas: Post-1948 Israeli Atlases and Their Successive Editions

With the establishment of a nation-state, the Atlas comes to represent its national story. Atlases narrate an origin story, portray political representations of a nation, and provide a space for "memory and systematic forgetfulness, fantastic allegories and geometric reason [to] coexist" (Rabasa in Sparke 1998, 480). New nation-states that were formed in the middle of the 20th century frequently used Atlases to tell their stories (Monmonier 1996). In Israel, too, the *Atlas of Israel* came to narrate a nationalistic story that set the standard for other cartographic products, ranging from geography schoolbooks to commercial maps. Its first edition, published by the Survey of Israel in 1957, was edited by one of the founders of academic geography in Israel: David Amiran. The first edition included dozens of physical, historical, political, and social maps of Israel that provided a detailed account of the history of a nation. Subsequent editions (published in 1956, 1970, 1985, 1995, and 2009, including an English edition in 1970) were not as comprehensive and were mainly to be used for educational purposes.

The *Atlas of Israel* became both an academic and a national project. While national agencies, such as the JNF, provided funding for the *Atlas*, Israeli academics were recruited to produce it. Academic trends, such as the focus on regional geography in the early 20th century, informed Israeli geographical approaches. The concept of the region and the emphasis on the symbiosis between a people and their physical, political-economic, and sociocultural environments informed the *Atlas*. Indeed, it is the symbiosis between the people and their environment that was the seed of national identity and national ideology. The *Atlas* became a national symbol and helped to promote the Zionist vision.

A comparison of the various editions of the *Atlas of Israel* reveals how the hegemonic construction (as well as some of the unintended counterhegemonic aspects) of a nationalist story was told over time (Sparke 1998). The *Atlas*'s narrative speaks to the fluidity of the notions of boundaries, the attempt of rooting the new state in a history and a region, and the fragility and smallness of the Jewish state within the Arab world. Already the name, *Atlas of Israel*, silences as much as it reveals. The Land of Israel differs from the State of Israel, thus does the *Atlas* represent the land or the state of Israel? It is only on opening the *Atlas* that we see that most of the maps relate to Eretz Israel (the Land of Israel) rather than the State of Israel. While the Land of Israel is defined within its stable and fixed Maimonides boundaries, the borders of the state are presented as provisional and temporary. Most maps portray the Land of Israel, which includes both sides of the Jordan River with no markings for political boundaries. Those maps that only relate to the State of Israel appertain to features in relation to relevant political boundaries at a particular time.

Indeed, according to an Israeli cartographer and former employee of the Survey of Israel, the Survey of Israel only includes boundaries that were relevant at the time of the maps' production. Accordingly, there is no need to integrate the Green Line in maps any longer. Not only was its cartographic elimination stipulated by a governmental decree post 1967, but also, in line with the official governmentally endorsed perspective, the cartographer maintains that the Green Line

> was the Armistice Line between Israel and Jordan. Since . . . there is a peace agreement between Israel and Jordan . . . there can't be an armistice line. . . . [A]n armistice line is a military line for an intermediate period, but after you have a peace agreement, there is no armistice line. So this is obsolete as an armistice line. (Interview 2011)

Accordingly, until 1967, the land was under Jordanian control and was subsequently de facto annexed by Israel. Israel's peace agreement with Jordan is therefore seen as having obliterated the status of the Green Line as an armistice line between Israel and Jordon. Yet, despite the official erasure of the Green Line from Israel's maps, it continued to be a line that matters. For instance, maps produced for regional planning exclude the West Bank, Gaza, and the Golan Heights, not least because these regions continue to be under military governance. Other maps that appertain to the physical features of the territory, however, include Greater Israel and embody the long Israeli tradition of branding a boundless Israel. As a result of the cartographic erasure of the Green Line, it has also been largely erased from Israelis' mental maps, despite its continued legal, administrative, and bureaucratic relevance (Fleishman and Solomon 2005; Portugali 1993).

While different types of maps may pertain to Eretz Israel and others to the State of Israel at any given time, how has the depiction of Israel's boundaries changed over time within the *Atlas*? A cartographic rendering of Jewish settlements during different time periods exemplifies how the *Atlas* tells the history of Israel in terms of the history of boundary-making (as illustrated in Figure 4.9).

While all four maps depict the areas west of the Jordan, they also include some territories on its eastern shores. On the first two maps from 1918 and

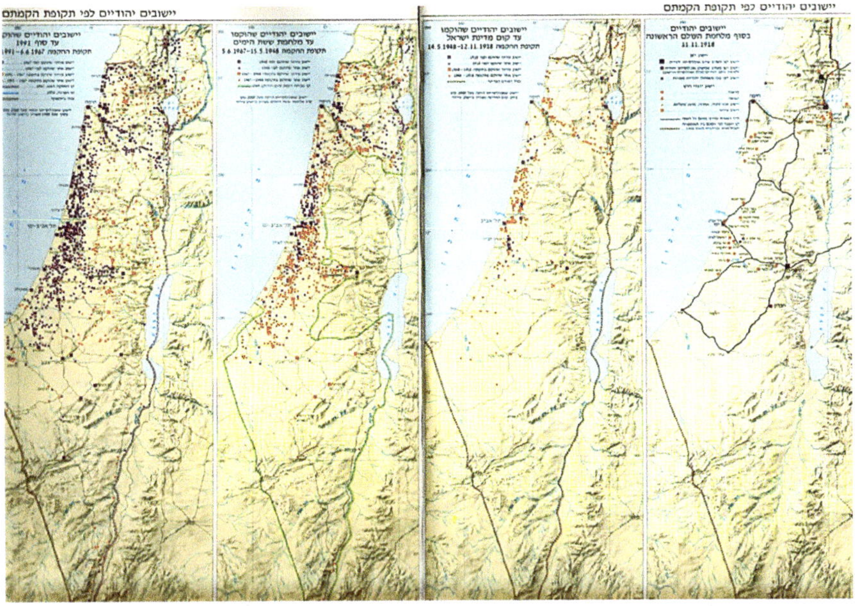

Figure 4.9. Jewish settlements by the period of establishment.
Source: Survey of Israel, *The New Atlas of Israel: The National Atlas* (Survey of Israel and the Hebrew University of Jerusalem, 2009), 76–77. Courtesy of Survey of Israel.

1948, respectively (if going from right to left), any boundaries, including those imposed by the British Mandate, have been eliminated. As mentioned previously, protagonists argue that the Survey of Israel depicts the boundaries relevant at any given time (Srebro 2011), yet when telling the cartographic history of settling the country, they have been erased. In this version of history, the increase in settlements takes place in seemingly empty frontiers that continued without the impediment of borders. In the third and fourth maps (again, from right to left), from 1967 and post-1967, respectively, the Green Line appears and then disappears. Yet, after 1967, while the Green Line ostensibly impacted the progressive settling of the country to the east, settlement activities focused on establishing a majority presence beyond the Arab-dominated "internal frontier" in the Galilee (Yiftachel 1991, 1996) as well as the desert frontier in the south. These maps portray the ethos of conquering the frontiers during different time periods (Kemp 1999; Kimmerling 1983a).

Ironically, the increasing use of satellite photos in later atlases and their juxtaposition with maps, lends the map, which is at the same scale as the satellite image, a sense of objectivity (as illustrated in Figure 4.10). And in doing so, this representational practice naturalizes the territory of the Land of Israel with a

Figure 4.10. From a Satellite Picture to the Map (Survey of Israel, Atlas of Israel, 2008), 9. Courtesy of Survey of Israel.

seemingly indisputable facticity. Juxtaposing the map with the satellite-rendered image of the land, however, also communicates an implicit counternarrative to the notion of a unified, Israeli-controlled territory.

The satellite image makes visible the different ecological, agricultural, and landscape practices on both sides of the Green Line. The landscape on the Israeli side appears darkish-green, in part as a consequence of JNF-funded afforestation projects inside Israel, while the Palestinian topology appears as light brown, due to overgrazing and desertification (Salameh 2008). In fact, the different grazing habits and environmental regulations in Israel and in the Palestinian Territories have become imprinted in the landscape and make visible the officially erased Green Line.

The *Atlas of Israel* focuses extensively on history. Many maps are either reproductions of historical maps, ranging from early maps, such as the historic Meidva map dating from the 6th century CE, depicting the hills, valleys, and villages of the Holy Land, to British Mandate maps produced in the 1940s. Various maps also depict historical narratives that emphasize the importance and long history of civilization in the Holy Land. A comparison of the first and last editions of the *Atlas* (1956 and 2009, respectively) reveals an emphasis on two different historical narratives across the two editions. Maps in the first edition focus primarily on ancient history, including what is considered to be the Jewish Golden Age in the region as well as the Israelites' revolt against the Romans. Only a few maps deal with other historical periods, such as the Crusades and Islamic rule over Palestine. The focus on Jewish history at the expense of other historical periods was part of early state-building efforts to which Israeli archaeologists, geographers, historians, politicians, and educators had been recruited. Their science and discourses could serve as ways to "naturalize" this contested land as a self-evidently Jewish land, justify Israel's historic right to that land, and implant in Israeli youth and new immigrants an attachment to the "homeland." On the other hand, in the 2009 edition, Jewish ancient history has been excluded in favor of Zionist efforts and their transformative power for land management. Historical justification for land claims has thus been superseded by the story of Zionists who make the land their own. The maps portray developments from the late 19th century to the present. Many of these cartographic renderings represent Jewish settlements as pioneering efforts to conquer the frontier, and they highlight Zionists' ability to make the "desert bloom" (as represented in Figure 4.11).

For instance, this series of maps tells a story of the progressive rationalization, modernization, and greening of Beit Shan between 1880 and 2006 (from right to left). The earliest map, produced by the Palestine Exploration Fund in 1880, shows the village of Beit Shan surrounded by swamps, marshlands, and farmland. In the 1937 map of the same area (produced by the Survey of Palestine), the swamps and marshland have been dried, with Beit Shan becoming a town set within an area of green spaces and fruit trees. The surrounding area, too, has experienced growth, with new Jewish villages and what were formerly represented

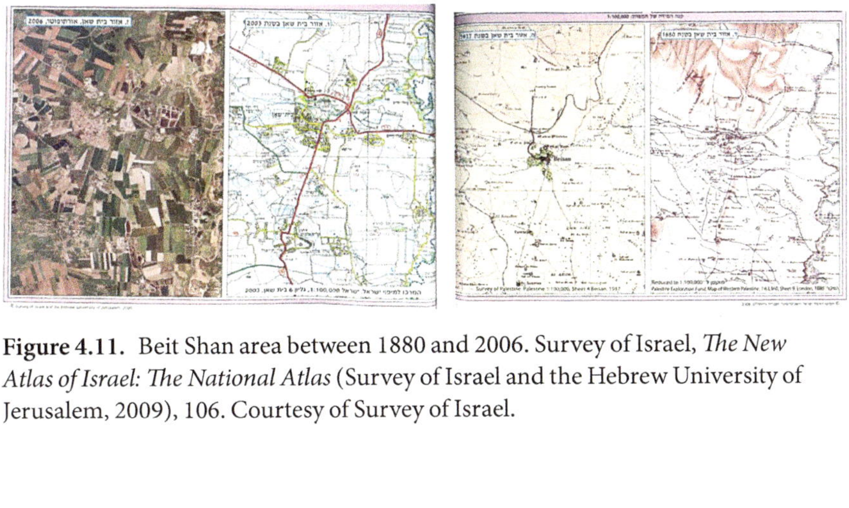

Figure 4.11. Beit Shan area between 1880 and 2006. Survey of Israel, *The New Atlas of Israel: The National Atlas* (Survey of Israel and the Hebrew University of Jerusalem, 2009), 106. Courtesy of Survey of Israel.

as an accumulation of Arab huts now permanent villages. The 2003 map (produced by the Survey of Israel) represents Beit Shan as a modern, planned, and highly developed town, surrounded by villages, fishing pools, cultivated lands, and an extensive road system. The 2006 air photo of the region emphasizes the culmination of all these efforts to develop the region: the Beit Shan valley has become a fertile green area that is geometrically arranged, rationally managed, and intensely cultivated.

This set of maps speaks to one of the myths of Zionism—its conquest of an untamed nature, marked by swamps, marshlands, and desert, and their transformation into a "Garden of Eden." Indeed the editors of the 2009 edition point out that they expanded the selection of maps so as to depict the fertilization of the desert as a distinguishing outcome of Jewish settlements (Survey of Israel 2009). Such efforts to settle and fertilize the land are ideologically sustained by religious as well as nationalistic sentiments. They are perpetuated by religious prophecies that proclaimed that the land would remain abandoned as long as the Israelites were in the diaspora. The swamps, therefore, were thought to have resulted from abandonment. At the same time, David Ben Gurion had also pronounced that fertilizing the desert and moving the desert frontier southward was to be one of the main goals of Zionism. Making the "desert bloom" thus became a source of national pride until the 1960s, and subsequently, the oPts became the new "frontier." The *Atlas* participates in these religious and nationalistic sentiments, sentiments that are fundamental to the Israeli nation-building project.

Last, the 2009 edition, unlike the first edition of the *Atlas*, also contains an array of maps that compare Israel to other countries around the world in terms of industrial and economic production, as well as sociodemographic data. The data presented show Israel (excluding the oPts) comparing favorably to other countries in terms of socioeconomic measures, positioning it among the most developed countries in the world. Together with the maps, they tell the story of a shift from a rural to a high-tech society. Maps celebrate a new Israel that is urban, capitalist, consumer-oriented, and yet environmentally aware. While the *Atlas*'s earlier edition focused on Israel's historical past and Zionist achievements, the 2009 edition emphasizes Israel's accomplishments as a developed country with a high standard of living. The story of Zionist redemption that seemingly rescued a land from backwardness and abandonment is thus celebrated as a modernizing engine in the heart of the Middle East.

However, one cartographic theme remained constant across different *Atlases* and maps: the notion of Israel as a small and crowded sliver of land in the Middle East (as portrayed in Figure 4.12a in which Israel is represented as a thin red sliver surrounded by various Arab counties depicted in green).

The theme of Israel as a vulnerable, densely populated, yet developed and productive territory that is smaller than most other nation-states, and at the same time surrounded by an Arab-dominated land mass, permeates various Israeli

Figure 4.12a. Israel Science and Technology Directory. Available at https://www.science.co.il/Arab-Israeli-conflict/. Retrieved May 28, 2019.

maps from the beginning of the Israeli state to the present day (see also Survey of Israel 2009, 128–133). Also, on a wall-size version of the first national map used in schools (as shown in Figure 4.8), a smaller map of Israel as part of the larger Middle East was placed to the side of the main national map. Israel (depicted by the color green) is shown as surrounded by Arab countries (represented by the complementary colors of purple and pink). The map emphasizes the smallness and vulnerability of the Jewish state within the seemingly homogenous and large Arab-controlled territories. Within the various Arab countries only few cities are marked. Such lack of cartographic detail reinforced the perception of Arab territories as consisting of empty spaces and wilderness that contrast starkly with Israel as a small, vulnerable, densely populated, fertile, and civilized territory.

Such maps of Israel as David facing the Goliath of the Arab World have always served as tools to legitimate the Israeli nation-state and its actions. Israel's smallness has indeed become a rhetorical trope. Consequently, many protagonists (usually associated with the Israeli right wing) argue that as "the Arabs" have 22 countries and the Jews one tiny state, surely, they should reach a territorial compromise with this state. Moreover, Israel's presumed vulnerability has also been a way to recruit people economically and militarily to defend the country and build up its capacity. Indeed, the myth of Masada reverberates through these cartographic depictions, as yet again, the Israeli state seemingly represents the last and only Jewish stronghold that will have to fight its enemies to its bitter end, and even now continues to be under siege by enemy forces (Bar-Tal and Antebi 1992).

The same types of maps that indicate Israel's smallness and consequent vulnerability are also produced by the Israeli Ministry of Foreign Affairs which publishes a range of maps of Israel superimposed onto land masses of other countries, such as England or the United States of America (as illustrated in Figure 4.12b).

Embedding the small land mass of Israel within much larger national territories is a visually effective tool to point to Israel's vulnerability. With official institutions promoting such cartographic themes, various social interest groups also increasingly use the same visual rhetoric to argue for Israel's right to expand its territory for security purposes and/or to accommodate Jewish migrants to Israel (a topic we will discuss in more depth in chapter 5).

The maps discussed so far were produced by governmental or quasi-governmental institutions. They represent the hegemonic construction of territories, borders, and boundaries as envisioned by the state at a given time, and bear the legitimacy of their institutions and have the allure of scientific objectivity. Official maps, whether they were designed for "propaganda" or "scientific" use, included depictions of an unbounded Land of Israel as well as a State of Israel within its relevant borders at the time. Arguably, it is what are taken to

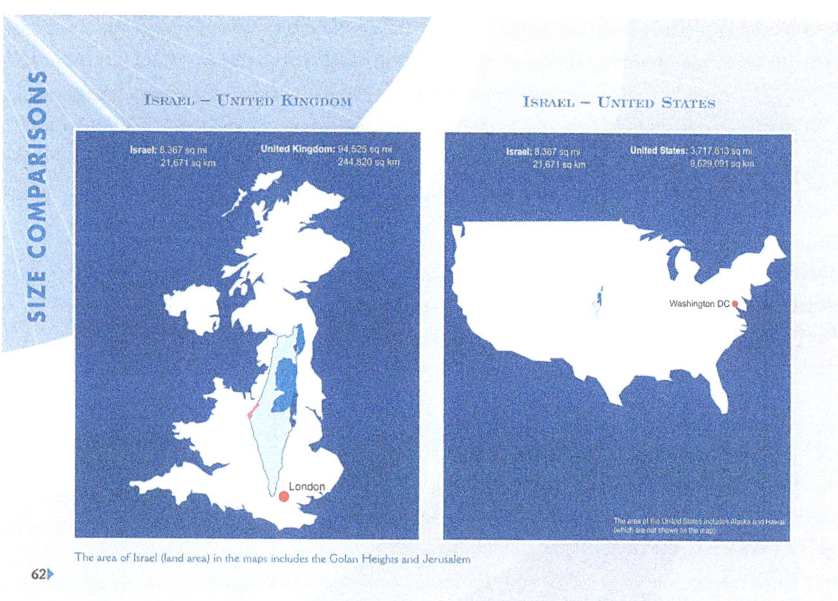

Figure 4.12b. Israel Size Comparisons. Israel Ministry of Foreign Affairs (2008). Available at https://mfa.gov.il/mfa/aboutisrael/maps/pages/israel%20size%20comparisons.aspx. Retrieved June 6, 2019.

be "scientific" portrayals of an unbounded Israel in national maps and atlases that become powerful visual tools for imparting a territorial imagination to a people. The power of what we take to be scientific maps lies in the very fact that their politics, unlike in "propaganda maps," is seemingly less explicit. Rather, political assumptions and values become part of the maps' subtext, often so deeply embedded within the image that they become taken-for-granted and naturalized visions of the territory (Giddens 1984).

Still, in Israel, as in other countries, not only governmental but also non-governmental institutions use maps to put forth particular territorial visions and social concerns. Some of their mapping practices overlap with officially enforced mapping conventions, yet others diverge. In the next section, we discuss the extent to which maps, produced by political parties up until 1967, overlap with or diverge from official governmentally produced maps in terms of their content and function.

Political Parties and Their Use of Propaganda Maps

Before and after the establishment of the Israeli state, political organizations and parties used both "propaganda" and "scientific" maps on posters and political leaflets to communicate political messages and territorial visions.[5] While both types of maps reduce, exaggerate, include, or exclude certain visual elements so as to maximize their visual impact, "propaganda" maps, more so than "scientific" maps, tend to authorize their knowledge claims less by adhering to cartographic standards and scientific objectivity and more by drawing on public discourses. To appeal to an audience, such maps tend to visually exaggerate and simplify a message, appeal to commonly held stereotypes and prejudices, and induce fear and hatred toward particularized others.

In pre-1967 Israel, the political landscape was ideologically split between the dominant social democratic political party Mapai (Mifleget Poalei Eretz Yisrael which was the Workers' Party of the Land of Israel) and the more radically left Marxist-Zionist party Mapam (Mifleget HaPoalim HaMeuhedet or United Workers Party). Other more marginalized political visions were put forth by the Israeli communist party at the time, Maki (short for HaMiflega HaKomunistit HaYisraelit), and Herut, the right-wing political party of the revisionist Zionists. While Mapai managed to largely establish a consensus on key issues including boundaries, Maki and Herut did not endorse the consensus. As a result, David

[5] Maps have long been used to further particular political visions. Already during both World War I and World War II, protagonists used the visual appeal of maps for propaganda purposes (Morton 2015; Barron 2015).

Ben-Gurion would famously proclaim in the Knesset "without Herut and Maki" in an attempt to move the consensus forward. Political parties such as Herut as well as Mapam, and Mapai frequently used maps to advocate for their political and nationalist vision.

Mapai was also known as the nation-building party. It had established the Israeli Defense Forces, pioneered the Jewish settlement projects, and provided the ideological foundations of the Zionist enterprise. It was less their socialist message than their nationalist sentiments that ultimately succeeded in establishing a consensus that enabled the social democrats to stay in power until their defeat by the right-wing party Likud in 1977 (Sternhell 1999). In the early 1950s, Mapai produced a political poster entitled "Growth under Siege" (as shown in Figure 4.13) that appealed to just such nationalist sentiments.

This political poster emphasizes the ostensibly risky and challenging conditions under which Mapai was building a nation-state. Not only was Israel faced with waves of immigrants and refugees, but pre-1967, it was also located on a sliver of land, and seemingly faced down by a stereotyped, faceless, menacing, and armed Arab "other." While confronted with these challenges, Israel (clearly illuminated and elevated from the rest of the lightly brown colored territory) is home to numerous urban and rural construction, settlement, and development projects that are part of the newly formed Jewish state. In contrast, the "empty" and "desert-like" Arab territories represented the perceived cultural homogeneity, economic underdevelopment, and ethnic danger of the surrounding Arab countries. Streams of Jewish immigrants arriving on Israel's shores underscored the need to build up the Jewish state, and at the same time, such population growth was seen as a way to counter the Arab threat. The map not only glorified the construction of settlements and industrial growth but also emphasized the spatial constraints on the Jewish project in a hostile environment. Indeed, this poster exemplifies a theme that informs cartographic representations of Israel in the Middle East up to the present: the "Arab threat." Mapai represented this as a demonized and stereotyped faceless Arab that endangers the constructive Jewish project. Political propaganda leaflets produced in more recent years by the right-wing Likud Party point to different threats emanating from the Palestinian Territories, ranging from Kathusyha rockets, enemy tanks, and terrorists. In 1980, the Embassy of Israel also published (as exemplified by Figure 4.14) maps that visually demonstrate the potential vulnerability of most of Israel to rockets launched from the West Bank, hereby indicating the indefensibility of the Green Line as a border.

Also in January 2009 the Samaria Settler's Committee produced a map illustrating that Tel Aviv was within rocket range, pointing out: "There are Jews in Judea and Samaria, there are no rockets in Tel Aviv" (Weiss 2009, 1). Such maps

Figure 4.13. The political poster "Growth under Siege" was produced by the political party Mapai in the early 1950s. Available at https://www.palestineposterproject.org/poster/growth-versus-siege. Retrieved June 6, 2019.

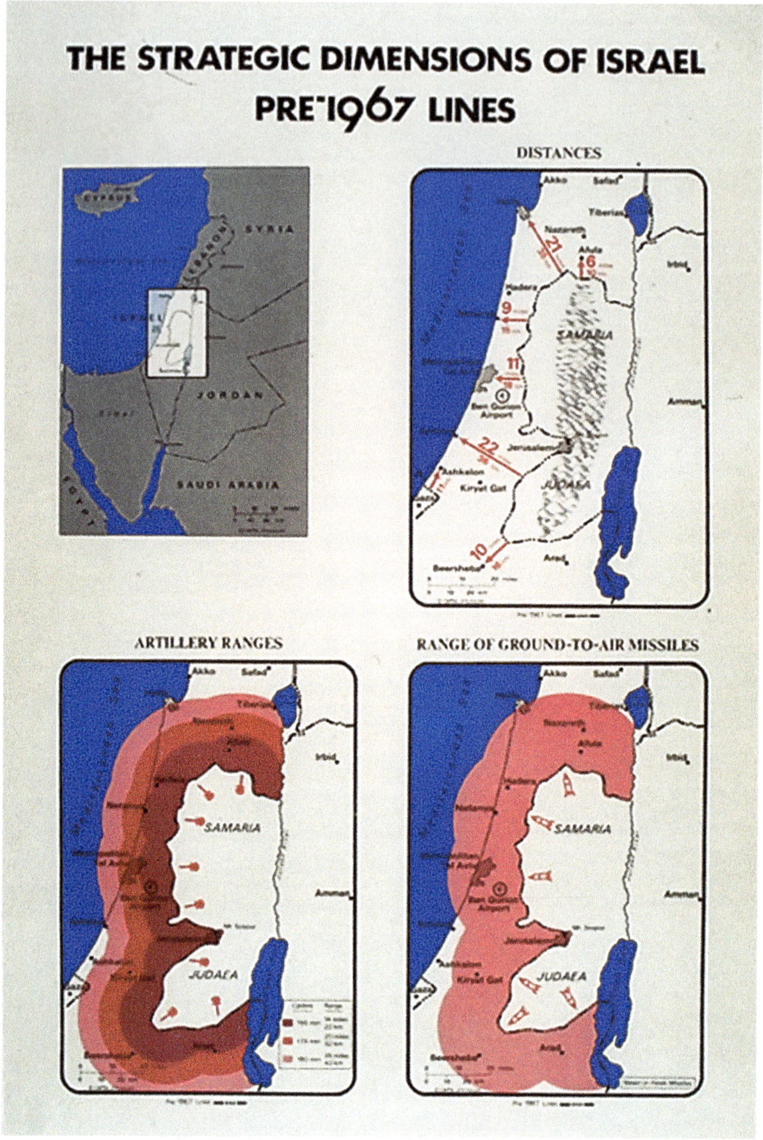

Figure 4.14. The Strategic Dimensions of Israel Pre-1967 Lines. Embassy of Israel, 1980. Available at https://www.palestineposterproject.org/poster/the-strategic-dimensions-of-israel. Retrieved May 28, 2019.

thus constitute a cartographic genre that continues to inform political discourses in Israel to this day.

The Marxist-Zionist party Mapam, which was seen as Mapai's more radically left-wing cousin, represented the Kibbutz movement. Mapam also used political posters to communicate ideological messages that were designed to represent and speak to rural settlers in collective settlements (such as Kibbutzim and Moshavim (a village that arose out of a cooperative farming community)) and to the urban working classes. Many of their posters celebrate the collective effort of nation-state building (as shown in Figures 4.15a and 4.15b).

The election poster produced by the Histadrut (the General Organization of Jewish Workers in Palestine) associated with Mapam is entitled: " 'M' —shield for the Homeland' [Magen Lamoledet]" (red and white enlarged text). The yellow text at the bottom spells out that "Mapam settlements are fortresses of strength for the country's borders" (as seen in Figure 4.15a). The poster portrays two young and determined settlers who, with guns in hand, settle and defend the Israeli frontier, work and cultivate the land, and build the nation.

Another Mapam poster produced by the Histadrut in 1954 (as shown in Figure 4.15b) emphasizes the importance of construction and defense for the survival of the Israeli socialist experiment. It portrays three men represent the working class in Palestine: a farmer (holding a shovel), a builder (holding a hammer), and a soldier (holding a rifle). The poster is textually framed by the biblical quote: "With one of his hands wrought in the work, and with the other held his weapon" (Nehemiah 4:17). The three men embody the national-socialist ideology as they cultivate the land, build its infrastructure, and defend its territory. It is these three tasks that were seen as crucial ingredients in building a nation-state, creating a new Hebrew culture, and forging a link between the new, strong, proud, pioneering, goal-oriented, and independent Sabra Jew (one who is born in Israel) and "his" homeland. Indeed, the Sabra negates the victimized diaspora Jew; instead "he" is rooted in the land and is determined to fulfill "his" mission of building a new society (Almog 2000; Schnell 1999). The new Sabra and Zionist is mostly male—it is men who are generally presented as having the mission to build the nation, while women may appear as the helping hands, and at times in charge of also defending their new homeland (Bernstein 1987; Herzog 1999). In these posters, workers and farmers—men and women—were working the land, while also defending it and, in the process, establishing the Zionist vision of a Jewish homeland.

As the maps above exemplify, the political posters produced by socialist parties, such as Mapai and Mapam, tended to delineate the boundaries relevant at the time. The Green Line often protruded from the map. Indeed, according to their socialist vision, it was not the territory and boundaries that were most crucial, but the creation of a new socialist nation. Their concern was less territorial

Figure 4.15a. Mapam Party poster from 1949. Available at https://www.palestineposterproject.org/poster/a-shield-for-the-motherland. Retrieved May 28, 2019.

than demographic—for the socialist project to succeed the state needed a solid Jewish majority dedicated to a common ideological vision of justice and equality. This socialist vision also helped shape the Labor Zionists' pragmatic attitude toward the issue of boundaries. In a similar vein, Ben Gurion stated at that time that territorial compromises do not preclude territorial changes in the future (Sternhell 1999).

Figure 4.15b. Mapam poster International Workers' Day 1954. Available at https://www.kedem-auctions.com/product/poster-international-workers-day-1954-shamir-brothers/. Retrieved May 28, 2019.

Unlike for the socialist parties, for the Zionist revisionist Herut Party, an independent state in Eretz Israel was the ultimate goal (Kellerman 1993). Their political posters included all the revisionists' claims to the land, including present-day Israel, the West Bank, Gaza, Jordan, and southern Lebanon. Indeed, Herut never recognized the British government's 1922 decision to separate Mandate Palestine from Transjordan, and thus never acknowledged the legitimacy of Jordan as a

Figure 4.16a. Revisionist Herut poster from 1948. Available at https://www.palestineposterproject.org/sites/default/files/Herut_1948.jpg. Retrieved June 6, 2019.

separate nation-state; instead, Herut continued to claim Israel's right to the whole of Eretz Israel.

The political posters presented here (as depicted in Figures 4.16a, 4.16b, and 4.16c) summarize Herut's expansionist territorial vision. In Figure 4.16a, the delineation of Eretz Israel is superimposed on Jerusalem's old city walls. The image of Jerusalem comes to signify everything that is central to Jewish identity and

Figure 4.16b. Revisionist Herut poster, "It Is All Ours," ca. 1935. Available at https://www.palestineposterproject.org/poster/all-ours. Retrieved May 28, 2019.

culture: Mount Zion, the Wailing Wall, and the Golden Age of Judaism. Unlike the Zionist Labor movement, whose members strove to create a new Hebrew culture and identity, Herut appealed to tradition, religion, and history. Also unlike in the Mapai or Mapam posters, in which it was not the territory but new human beings and their socialist vision for a new state that were foregrounded, Herut clearly cartographically delineates its land claims on Eretz Israel. In this poster, this nationalist message is textually enforced with the statement: "The

Figure 4.16c. Herut poster, "Only So," ca. 1935. Available at https://www.palestineposterproject.org/poster/sole-solution. Retrieved May 28, 2019.

walls of ancient Jerusalem are not the boundaries of Jerusalem. The Jordan is not the boundary of our country and the sea is not the boundary of our nation."

Other Herut posters expressed these territorial claims most explicitly (see Figure 14b). Inside a demarcation of Greater Israel, the text states: "It is all ours," and framing Eretz Israel is the well-known Zionist slogan: "It is good to die for our land." Such visually abstract and uncluttered visual and textual statements were to persist in a political climate that, at the time, was dominated by the Labor

Zionists who had compromised on Israel's boundaries. Herut's provocative visual language thus went hand in hand with the radicalness of their claims at the time. Their marginalization from the political mainstream, which was unchecked by the need for political compromise thus also shaped the visual and textual messages they attempted to communicate to the Jewish public.

The poster "Only so" (in Figure 4.16c) portrays Herut's logo and ideology most succinctly. The logo consists of a gun in hand superimposed on Greater Israel; this logo is textually framed with the words "Only So." Accordingly, only the gun will provide the Jewish people with Greater Israel. Indeed Ze'ev Jabotinski, one of the movement's founding leaders, had in 1925 called upon Jews to immigrate, conquer, and control the country with the might of a Jewish army (Naor 2001). He objected to the Labor Zionists' vision of making the land their own by settling and working it, claiming that only with guns in hand will Greater Israel became theirs. For Herut then, control over territory was a way to give refuge to Jewish people in the diaspora and provide a state that would represent their interests.

As we have seen, in pre- and post-1948 Israel, maps became a crucial tool for forging a nation-state, and re-naming the land, rivers, and towns was a way to control the territory and make it one's own. Re-naming the territory also went hand in hand with constructing its history by emphasizing some historical periods over others. These efforts informed official mapping efforts as exemplified by the making of the *Atlas of Israel*. The *Atlas* delineated a country, re-named its landscapes, and told its history. It becomes, as elsewhere, a symbol of a nation. At the same time, political parties, inspired by the visual power of maps, increasingly used maps to communicate their ideological concerns and territorial claims. The nation-building party of the Labor Zionists Mapai was pitted against the more radical Mapam Party, the communist Maki Party and the revisionist Zionist Herut Party. Their different visions for a new society, a new human being, and a new state commanding a yet to be defined territory became part of a cartographic debate that still reverberates in Israeli public discourse to this day.

5
Maps with a Mission Post-1967
Securing Jewish Land

Introduction: Cartographic Wars over Israel's National Territory Post-1967

At the end of the 1967 war between Israel and its neighboring states Egypt, Jordan, and Syria, Israel had more than tripled the territory under its control, occupying the West Bank, East Jerusalem, Gaza, the Sinai Peninsula, and the Golan Heights. While under the Labor Zionists there was largely a consensus over the contours of Israel's national territory, after Israel's territorial expansion the struggles intensified greatly between the "hardliners," who advocated territorial expansion and still dreamed of a Greater Israel, and the "softliners," who wanted a territorial compromise with the Palestinians (Kimmerling 1983b, 1998). In 1967, Labor Minister of Education Yigal Alon added fuel to the fire when he implemented a Knesset decision to erase the Green Line from Israel's official maps. With this decision the relative consensus concerning boundaries during the pre-1967 period had evaporated. Instead, the action instigated an all-out war of maps over how to delineate Israel's national boundaries. Particularly since the 1980s, such map wars were facilitated by the rise of new mapping technologies and software that put the tools of map-making into every home and office with access to a computer and the internet. Like elsewhere, where the rise of the increasing democratization and privatization of mapping practices had increasingly transformed cartography from a disciplinary expertise to a people's cartography, in Israel too, maps became an ever more widely used visualizing technology and yet another way to swing the sword (Radcliffe 2009; Schnell and Leuenberger 2014). Political interest groups and NGOs from the Israeli right and left increasingly capitalized on the ambiguities of Israel's newly revised borders to put forth their political and territorial visions of the territory, boundaries, and ethnic identity of the land.

Once the Knesset decided to eliminate the Green Line after the 1967 war, cartographers had to follow suit. As the leading Israeli cartographer points out:

> On a map which I published for schools . . . we did not put in the armistice line. . . . they did not want to recognize the boundary. It was a policy of the

The Politics of Maps. Christine Leuenberger and Izhak Schnell, Oxford University Press (2020).

DOI: 10.1093/oso/9780190076238.001.0001

> government not to recognize [it]. This is just a kind of an administrative line, because the sovereignty of the whole area was Israeli and there was pressure also from the right-wing elements not to put it. . . . [M]aps that go to schools and the public are printed by the Survey of Israel—a government institution, and they refused to publish it. (Interview B 2008)
>
> Consequently there were "usually no marking or mention of the pre-1967 border, which was expunged from the maps" (Collins-Kreiner, Mansfeld, and Kliot 2006, 395). While before 1967, the Governmental Names Committee's mandate was only for Israeli territory, after 1967 the newly acquired Israeli-controlled territories were also Hebraized in order to produce toponymical consistency on both sides. The committee was also called upon to provide names for Jewish settlements in what, according to International Law, was now the oPt.

Disputes about Israel's post-1967 boundaries nevertheless endured within Israel. Indeed according to one cartographer, some Israeli organizations:

> acquired the maps . . . and took [them] to another press . . . [which] put in the Green Line. . . . They wanted people to see exactly what is Israel, according to their political view. They don't want to annex. . . . [T]hey are against settlements in the West Bank. (Interview B 2008)

Despite the fact that post-1967 Israel had annexed East Jerusalem and made Israeli law (for settlers) and martial law (for Palestinians) applicable to the West Bank and Gaza, controversies over their territorial status nevertheless persisted. These territories retained a separate standing not only because of political controversies over their status and inconsistent Israeli government policies, but also because Palestinians retained their own semiotics of the land (Feige 1999). For some Israeli human rights organizations, such as B'Tselem, the fact that acquiring land by force, as well as occupying and settling it, is considered illegal under international law also helped retain the disputed status of the territories.[1] Consequently, cartographic demarcations of Israel's borders are tied more to mapmakers' whims than to border markers on the ground. Moreover, as maps became ever more politicized, geographers and politicians often found themselves embroiled in public controversies. With the consensus over the limits

[1] B'Tselem, unlike some other Israeli and Palestinian organizations, is aligned with United Nations mapping conventions, by recognizing and drawing only those borders that are based on mutually agreed upon peace deals. Due to the lack of an Israeli-Palestinian peace agreement concerning borders, the 1949 armistice line (the Green Line) therefore demarks the two territories instead.

of the Israeli national territory all but evaporated after 1967, various political sandstorms over how to map the newly occupied territories in the West Bank and Gaza ensued.

One notable public outcry was sparked by the publication of the third edition of the *Atlas of Israel*, edited by David Amiran (Amiran 1985). In text accompanying the maps, Amiran implied that the locations of Jewish settlements were determined by political rather than geographical considerations. After a journalist drew attention to Amiran's statement, the right-wing Likud minister David Levi stopped publication of the *Atlas* (see Bar-Gal 2004). According to an interviewee, "he [Levi] decided that the version was contrary to Israeli interests, was insulting to the ideology of the Israeli government and as the *Atlas of Israel* is an official publication of the government . . . a page had to be removed from the *Atlas*" (Interview B 2008). Israeli geographers, regardless of their political views, were united in their protests and refused to rewrite the section on the settlements, as doing so would have challenged the academic integrity of the *Atlas*'s editor. The *Atlas* was eventually published, but with an insert solicited by the Israeli housing minister from a non-geographer, whose statement legitimized the settlements. This debate was not over the map's visual elements but over the textual signifiers used to critique or justify the location of Jewish settlements within the West Bank.

A second uproar occurred when Moshe Brawer, a leading cartographer from Tel Aviv University, published a set of editions of the world atlas for schools. Although the atlas followed the post-1967 convention of eliminating the Green Line and using Hebrew names for the region, the minister of education from 2001 to 2006, Likud party member Limor Livnat, excluded the atlas from schools since it represented the West Bank (also known within Israel as Judea and Samaria which are considered to be part of the Israeli national space that is made up of Judea, Samaria, Benjamin, and Gaza) as distinct territories separate from the state of Israel. As one interviewee explained:

> One day the council of the settlement in the West Bank wrote a letter to her [Livnat] that in the Atlas, in maps for school, they are not in the map. . . . [T]he manner in which the West Bank is presented gives the impression that it is a separate political identity, and not part of Israel. So, based on this the Minister instructed the director of the Ministry of Education to forbid the use of the Atlas in schools (Interview B 2008).

Meanwhile, at the same time newspaper headlines in a leading Israeli newspaper, *Yedioth Ahronoth*, proclaimed that "an Israeli geographer has already recognized the West Bank as an independent state" (Interview B 2008).

Despite Brawer's academic standing (he had received the prestigious Israel prize in 2002)[2] and protests, Limor Livnat insisted that the *Atlas* could only be published after changes were made. Brawer, under pressure from the publisher and Livnat, "put the West Bank in a frame together with Israel" and he agreed to print the headline over Judea and Samaria not "in the same bold letters as Israel but much smaller." Consequently, "the whole set, that is Israel, the West Bank, and Gaza strip will be enclosed in a black frame so the pupils will know this is one unit" (Interview B 2008). In this case Brawer and Livnat disagreed over both visual and textual signifiers. How to visually represent Israel and the Palestinian territories, and how to use headings to signify Judea and Samaria as being subordinate to the state of Israel could both, in different ways, indicate either the unity or the separateness of these territories.

A third dispute involved a Labor minister of education, Yuli Tamir, who was also a former Peace Now activist. In 2006, Tamir decided that the Green Line and the pre-1948 British Mandate border should be added to the maps for schools. She argued that Israelis couldn't condemn Palestinian maps for not recognizing Israel when Israeli-made maps had cartographically eliminated the Palestinian territories. Moreover, for her, the history and politics of the region could only be appreciated by understanding the past delineations of boundaries. Indeed, she maintained that for students to learn about the 1967 war and the resulting boundary disputes is fundamental if they are to understand current Israeli-Palestinian politics and the controversy over the UN Security Council Resolution 242 that passed on November 22, 1967, calling for an end to the conflict and for the withdrawal of Israeli security forces from the oPts[3]. After all, before students could engage in political debate, they had to learn the facts (Medzini 2011). This, however, sparked a political storm. The settlers' council responded by attacking the minister for promoting what they referred to as her personal political agenda, alleging that she was not motivated by educational or scientific rationales but by her desire to indoctrinate children with a left-wing version of the national map. The settlers' rabbis disallowed the use of textbooks that included the Green Line and threatened to boycott any texts authorized by the Ministry of Education that did include the Green Line.

2 "The Israel prize" is Israel's most prestigious prize given out annually on Israel's Independence Day. It is bestowed upon distinguished scholars, artists, and people for their contribution to society. According to the prize committee, Brawer distinguished himself by producing atlases, pioneering the study of Arab villages, and founding geography departments at Tel Aviv and Bar-Ilan University.

3 For more information on UN Security Council Resolution 242, see https://unispal.un.org/DPA/DPR/unispal.nsf/0/7D35E1F729DF491C85256EE700686136. Retrieved January 25, 2020.

Benjamin Netanyahu (long-term chairman of the Likud Party and Israeli prime minister) also pointed out that:

> over the years, I and others have been fighting to change the marking on the map so as not to determine the results of future agreements, and now Tamir comes in and decides the marking should include the Arab demands. (Rettig and Frenkel 2006, 1)

Between this wave of public criticism and the right-wing onslaught, the minister's attempt to re-introduce the Green Line on Israel's map did not survive the political storm it produced.

These examples demonstrate the myriad ways that Israeli cartographers and advocates for particular cartographic visions could quickly find themselves embroiled in political controversy over how maps should represent the status of various territories. Given the lack of consensus over the territorial shape of post-1967 Israel, governmental and non-governmental institutions, ranging from right-wing interest groups to human rights and peace organizations, have been at the forefront of these territorial debates. They frequently rely on maps as tools to constitute the national territory, its boundaries, and alleged ethnic identity. These organizations use diverse cartographic strategies to relay their concerns. Some use a mixture of textual and visual signifiers as well as spatial demarcations to make maps seem accurate, objective, and scientific whereas other map-makers design maps that lay no claim to scientific accuracy and often communicate emotional messages. And it is to these we turn now, focusing in particular on how, after 1967, political parties, such as the right-wing Likud Party; official bodies, such as the settlers' council and regional municipalities; commercial mapping companies, that produce tourist maps; and the right-wing neo-Zionist organization MyIsrael[4] have partaken in such map wars.

Maps with a Mission: Securing the Land, the Bible Land, and Jewish Land

Together these interest groups and organizations represent a range of generic mapping practices that are informed by a number of commonly shared discourses and concerns. Their maps communicate different messages, ranging from concerns about Jewish historical and territorial rootedness in Greater Israel to Israel's national security. We now turn first to the ways that Likud, the Yesha Council (which represents the interests of the Jewish settlers within the oPt[5]),

[4] See MyIsrael, http://www.myisrael.org.il/action/en/.
[5] See the Yesha Council, http://theyeshacouncil.org/.

and MyIsrael exemplify three commonly shared discourses associated with the Israeli right wing: the "secular expansionist" discourse, which advocates a territorially broad Israel for security and defense purposes; the "national religious" discourse; and the "Jewish rootedness in the land" discourse. The latter two both encourage Jewish settlements in Greater Israel and tend to hebraize and "judaize" what is considered by their adherents to be historically Jewish land.

With the newly acquired territories to the east after 1967, Israeli right-wing parties (who in the past had been subdued under the Zionist Labor consensus) produced a set of political propaganda maps to put forth their long-standing visions of a Greater Israel. The Zionist Labor Party and other left-wing parties, however, who had long accepted the Green Line as Israel's practical boundary and who were in favor of territorial compromise with the Palestinians, stopped distributing maps depicting Israel and its borders. They did not want to add fuel to the fire and alienate potential party supporters, especially those who might have settled in the newly acquired territories in some of its many Jewish settlements. Even mention of the Green Line and the possibility of reinstating the more limited pre-1967 borders by delineating them in maps become a political red flag.

With war won and territory expanded, the 1970s saw the growth of right-wing ideologies. These can be divided into a "secular expansionist" camp and a "national religious" camp. The center-right party Likud is an example of the "secular expansionist" camp. Likud was founded in 1973 and its ideological goals have long included supporting free market capitalism and Jewish settlements in the West Bank (Judea and Samaria), while opposing Palestinian statehood. Likud came to power with the 1977 elections. This was a major political shift as it was the first time in Israel's history that the left lost power.

Securing the Land

During the 1970s and 1980s, Likud frequently produced maps as part of its continual political campaign against any peace initiative that involved withdrawing from the oPt. A generic feature of these maps is the omission of cartographic details and their highly abstract delineations of the territory. Such cartographic minimalism can produce a powerful univocal message that is easily accessible to its audience. These cartographic depictions were also frequently emotionally charged and often appealed to the public's fear of the "Arab threat" as they emphasized either the threat to Israel's national security, should it withdraw to the 1949 armistice line, or highlighted Israel's smallness and

Figure 5.1. Likud Party's Youth Movement map against Shimon Peres.
Source: Courtesy of the Jabotinsky Institute in Israel.

vulnerability. This "secular expansionist" rhetoric, that focused on issues of national security, defense, and outside threats, became a particularly compelling tool to reach across party lines and gain support against withdrawing from the oPt. For instance, the Likud Party's Youth Movement[6] actively promoted their political visions through propaganda maps (as shown in Figure 5.1).

[6] See Likud Party Youth Movement, https://www.likud.org.il/en/the-likud-youth-movement.

In this map, the contours of Israel are made up of oil pipes with its "narrow soft belly" tightened by a constricting watchstrap. The text accuses the then Labor candidate Shimon Peres of having received personal benefits from an oil deal with Egyptian companies, including a watch, in exchange for the establishment of a Palestinian state in the oPt. The price of Peres's dealings and visions are thus presented as being Israel's territorial loss: the return of the occupied territories to the Palestinians.

By the 1981 political election Likud produced an array of maps in its struggle for political power by communicating its firm commitment to retaining the Palestinian Territories and opposing any territorial concessions. A political poster with the words (in large bold red letters) "the Likud will prevent this!" and (underneath in smaller black letters) "2.5 million Israelis are within firing range of enemy artillery if an 'Arafat state' is founded," exemplifies the visual and textual means by which Likud presented itself as the guardian of Israel's security (as displayed in Figure 5.2).

The map sketches only the general silhouette of the land while emphasizing the security risks emanating from the West Bank should Israel choose to withdraw to 1967 borders. The four red radiuses depict the threat of "Arafat artillery" to the central plains of Israel. Major population centers from north to south—from Afula, Netanya, Tel Aviv, Jerusalem (which despite its disputed territorial status since 1967 is presented as part of Israeli territory), Ashdod, and Beer Sheva to Arad—are all depicted as within the range of artillery fire. The use of the name "Arafat" (who was largely demonized within the Israeli "secular expansionist" camp) to describe a Palestinian state aims to strengthen public fears about a possible withdrawal to 1967 borders. The map thus uses visual and textual signifiers as well as spatial ordering to convey Likud's vision of the West Bank. The color red, combined with the tanks stationed across the seemingly large and visually dominant West Bank territory connotes a severe threat to the narrow strip of densely populated Israeli territory.

Another map produced at that time, entitled "An Arafat State Will Not Arise," shows the short distances between the Green Line and the Mediterranean Sea and consequently between the "Arafat State" and major Israeli populations centers (as indicated in Figure 5.3).

The map is spatially demarcated so as to emphasize the size and potential threat to Israel of a Palestinian-controlled West Bank. It highlights metropolitan Tel Aviv as Israel's core, with red arrows stressing its proximity to the Green Line. Other major Israeli population centers, from Beer Sheva to Afula, are also shown within easy reach of the West Bank. The distances between the Green Line and Israel's Mediterranean shoreline appear alarmingly short, while the horizontal and vertical measurements projected across the West Bank make this territory seem overly large. Jerusalem is depicted as Israeli territory and is

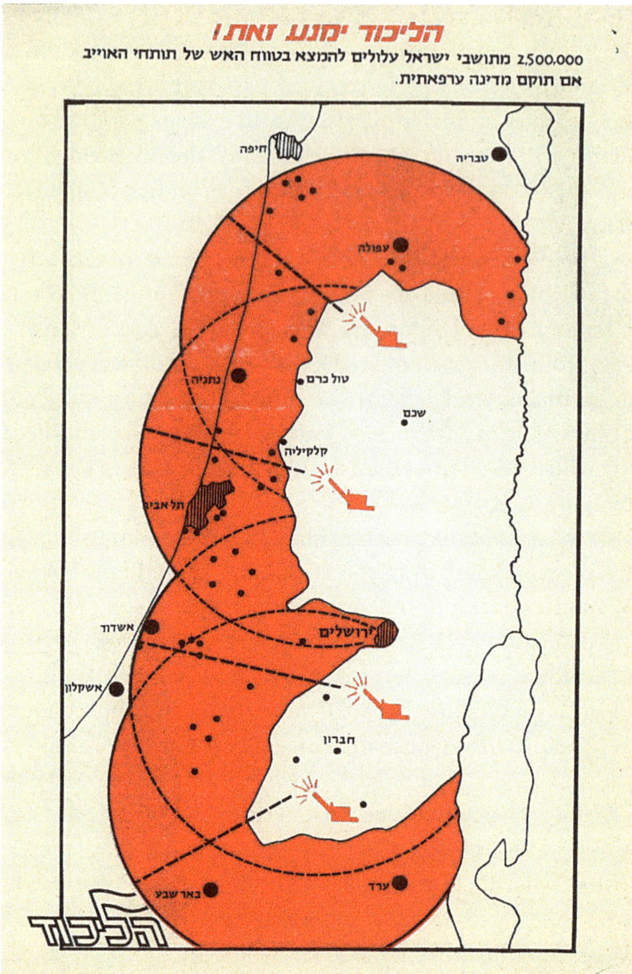

Figure 5.2. Likud Party's political poster "The Likud Will Prevent This!" Courtesy of the Jabotinsky Institute in Israel.

shown as being especially vulnerable given its proximity to the West Bank. Here the map's heading and the red arrows produce a powerful mix of textual and visual signifiers expressing Likud's opposition to a possible Israeli withdrawal to the Green Line on security grounds. The political position expressed in this map (presented in Figure 5.3) continues to represent Likud's stance. As World Likud chairman Danny Danon pointed out in 2006, returning to the Green Line would be dangerous "to the existence of the State of Israel and . . . would turn us all into hostages in the hands of Palestinian terrorists" (Efrat Weiss 2006). Such

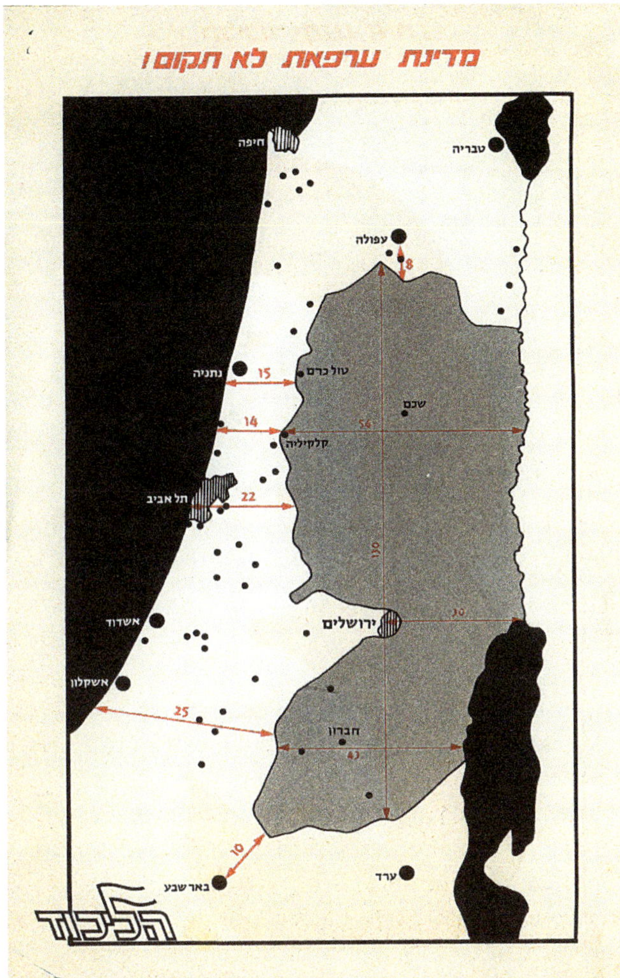

Figure 5.3. Likud Party's political poster "An Arafat State Will Not Arise." Courtesy of the Jabotinsky Institute in Israel.

Likud maps thus continue to epitomize the cartographic genre that depicts Israel as vulnerable and under threat as discussed in chapter 4.

During election campaigns, Likud tended to produce an array of provocative political posters, but such political point scoring subsided with its ascent to power. Once Likud found itself on center stage politically, the party adopted a more pragmatic and compromising attitude (Ben-Rafael Galanti, Aaronson, and Schnell 2001). Without changing their commitment to their claim for a Greater Israel, at a Likud Party convention in the early 1990s, they

Figure 5.4. The Likud symbol for the convention once in power.
Source: O. Potchter and O. Gravis Kovalski. 2011. "The Map as a Political Symbol." In I. Schnell and E. Lavie, eds. *The Political Role of Maps in the Israeli-Palestinian Conflict*. Steinmetz Center for Peace: Tel Aviv University, Tel Aviv.

chose to have a banner hide the territorial controversy, with the whereabouts of the boundaries deeply buried under it (as exhibited in Figure 5.4). While the seemingly territorially plump Israel creates the visual impression of Likud preserving a maximalist territorial attitude, the banner spelling out Likud simultaneously hides any territorial concessions that might be made, even under a Likud government.

In 1993 the Oslo Accords were ratified. They constituted a set of interim agreements between the Israeli government and the PLO creating the Palestinian Authorities, which was tasked with limited self-governance of parts of the West Bank and Gaza. A final status agreement was to be negotiated within five years. The Oslo Accords spatially divided the Palestinian Territories into Zones A, B, and C, representing different legal territorial statuses. Zone A denotes full Palestinian responsibility over the territory, Zone B is administered by the Israeli military and the Palestinian civil administration, and Zone C is under full Israeli military control. At the time, fears of a Palestinian demographic threat (Soffer 1989) increasingly led to the recognition that some territorial compromises were in Israel's interest as otherwise the higher Palestinian fertility rate could quickly threaten a Jewish majority state. During this period, Likud and other political parties largely refrained from publishing maps. Concerns increased that published maps could erode a party's political base in areas (mainly inhabited by settlers) that were not included within Israeli territory. Also, politicians were aware that maps could become the basis for negotiation and compromise over land claims with the Palestinians, and no political party wanted to be accused of producing maps that failed to include territories claimed by various Israeli interest groups. Given the political implications of where these borders might lie, parties across the political spectrum preferred to keep the borders amorphous and stopped producing maps for political campaigns. While the question of boundaries all but disappeared from official political discourse, they resurfaced again all the more forcefully among politically concerned NGOs from the right and the left.

While Likud has produced some of the more visually powerful maps articulating the position of the "secular expansionist" camp, religious organizations (such as the Third Temple), governmental agencies, political interest groups, and settler lobbies have also created maps pointing to the smallness and vulnerability of Israel and its needs for security and self-defense. The Yesha Council, in one map, relied on similar cartographic tropes to depict Israel's vulnerability and smallness by providing an abstract topographical map that highlighted the mountainous regions within the West Bank that allegedly tower over the low-lying plains of Israel and hereby present a security risk. Another political interest group, the American Israel Public Affairs Committee (AIPAC), which aims to secure Israel's future, lobbies for Israel's interests on Capitol Hill in Washington, DC, and aims to strengthen US-Israel collaborative ties; it also publishes a range of online maps with similar objectives. These maps visualize such concerns as Israel's smallness and its need for protection, the threat of terrorism across its borders, and "Israel's narrow waistband," which "measures a mere nine miles" between the edge of the Mediterranean Sea near Natanya and the Green Line.[7]

Such maps from the "secular expansionist" camp (as exemplified by Figures 5.2–5.4), whether they stem from Likud, the Third Temple, the Yesha Council, or AIPAC, exhibit some of the cartographic strategies used to create, increase, and secure what are considered to be Jewish national spaces. They tend to use various visual and textual signifiers as well as the demarcation of space to point to Israel's vulnerability within a seemingly hostile Middle East. When stressing security concerns, these maps do not appeal to cartographic exactness. Rather, the "secular expansionist" camp demarcates the space cartographically so as to highlight an "Arab threat" which supports the need to expand Israeli territories for reasons of security and defense. The argument for expanding Israeli territory for security purposes goes hand in glove with the desire of the "national religious" camp to establish a Greater Israel based on biblical prophecies to which we now turn.

[7] See maps from American Israel Public Affairs Committee at https://www.aipac.org/. Retrieved April 26, 2020.

The Land of the Bible

The ideological underpinning of the "national-religious camp" was inspired by the political movement Gush Emunim, which arose in response to the 1967 and 1973[8] wars and was formally established in 1974. Gush Emunim actively encouraged Jewish settlement of Greater Israel as a religious duty while also frequently invoking the security and defense discourse (associated with secular right-wing expansionist views) in order to justify such settlements. It became the most successful right-wing religious movement within Israel, impacting both public perception and governmental policies. By expanding its networks into schools, synagogues, and religious seminaries (yeshivot), it became a powerful "extra-parliamentary mode of protest" (Newman 2005, 195) that brought about various settlements and political and ideological organizations. While Gush Emunim formally ceased to exist in the 1980s, its sentiments and ideological tenets underwent a process of governmental cooptation and institutionalization, which only increased the popular appeal of the settlement movement.

The powerful settler's lobby, the Yesha Council, grew out of Gush Emunim. It was founded in the 1970s and to this day serves as an umbrella organization representing the interests of Jewish municipalities within Judea and Samaria (the oPt). The council is committed to promoting settlements in Eretz Israel as a way to return to the biblical homeland, and it lobbies for settlers with the Israeli government. Maps produced by such organizations tend to emphasize Jewish rootedness in their biblical homeland, attend to the Hebrew topography of the land, and cartographically construct the unity and inseparability of the Greater Israeli territories. For instance, the Yesha Council produces a tourist map entitled "Heart of the Bible Land" (as shown in Figures 5.5a and 5.5b).

The title implies that this region is living testimony to the biblical period and thus the heart of the land. The depiction of ancient ruins, domes, and symbols as well as the wild untamed nature on the cover page convey the area as a biblical land untouched by secularization, modernization, and political conflict. The depiction of the jeep indicates that the land and its exotic sites are a safari experience accessible only with a four-wheel-drive vehicle and a sense of adventure. Judea and Samaria are thus both a frontier and also a historic treasure hunt (Sheleg 2000; Gurevitz 2007).

The map itself (as represented in Figure 5.5b) employs various visual and textual signifiers to stress Jewish rootedness in Greater Israel and the Jewish identity of the space represented. Settlements appear large and are demarked by the

[8] The 1973 war, also known as the Yom Kippur War, was a war by a coalition of Arab states led by Egypt and Syria against Israel.

Figure 5.5a. Tourist Map Lev Ha-aretz Heart of the Bible Land.
Source: Yesha Council.

Zionist symbol of the white, red-roofed houses. They are shown in relation to archaeological sites from the biblical period, which signify early Jewish presence in the Holy Land. These sites and settlements are also seemingly well connected by the visually prominent Israeli road system. While with the rise of modern cartography, 19th-century conventions of using ornamentation, a lack of scale, and enlargement of "important sites" when depicting the Holy Land had ostensibly passed, these maps attest to their current use and the intermittent inroads of scientific cartography into mapping practices.

Additionally, in an effort to point to Jewish historical sites and construct a biblical map of the land, various Arab historical sites were renamed using biblical and Jewish terms (despite lack of scientific support within the Israeli archaeological community for renaming these sites) (Feige 1999). Also, while Palestinian-controlled areas (Zones A and B) are indicated through shady yellow areas, Palestinian towns and villages are absent from the map; hence the whole geographical space is presented as being Jewish. Only on the cover page do the producers imply the separate status of the West Bank by expressing their gratitude

Figure 5.5b. Tourist Map Lev Ha-aretz.
Source: Yesha Council.

to the Israeli military, reminding drivers that travel is secure and approved along the routes of this biblical land.

Other regional maps (e.g., the Hebron mountain map produced by the regional municipality of Judea) display similar characteristics as the maps cited above. They often use Hebrew place names and omit Arab names for towns, villages, and streets. For instance, the regional municipality of Judea created a map of the city of Hebron, in which there is one Jewish settlement within a predominantly Palestinian city. The map provides Hebrew names for the city streets (including those in Palestinian areas) in an attempt to transform the identity of the city. Such maps mainly include Jewish settlements (larger Palestinian population centers may be marked, but frequently in smaller letters than smaller neighboring Jewish towns), emphasize Jewish archaeological sites, and aim to attract settlers to the area. They characteristically represent Judea and Samaria as part of Israel, as can be seen in a brochure produced by the regional municipality of Samaria, entitled "The Ma'ale-Shomron Hike," which shows Jewish towns and settlements on both sides of the missing Green Line. As David Newman points out,

> the naming of places . . . plays a major role in the creation of identity landscapes. Imagined Biblical territories of the past are transformed into concrete, living territories of the present, thus forging the link between historical myth and contemporary political reality. . . . Palestinian villages . . . have been replaced by Israeli-Jewish communities bearing Biblical or modern Zionist names in their place. (Newman 2001, 239)

A tourist map of the Mount Hebron region exemplifies the Israelizing of the region most starkly. It depicts larger Jewish cities within the Green Line, such as Jerusalem, Beer Sheba, Kiryat Gat, and Arad and inflates the size of small Jewish settlements within the West Bank such as Kiryat Arba (located within the Palestinian city of Hebron, or El-Halil), depicting them as relatively large settlements, with Hebron the only principally Arab city mentioned, and solely because it contains the Jewish settlement of Kiryat Arba. Such presentational styles stress the connectivity of West Bank settlements with major Israeli cities, whereby geographical spaces on both sides of the Green Line seemingly become one continuous and unified space. The cartographic erasure of the Palestinian population, the emphasis on Jewish archaeological sites, and the Hebraization of settlements jointly create an imaginary Jewish space that extends over both sides of the missing Green Line.

Such Israeli-produced maps use visual and textual signifiers to turn a culturally and ethnically heterogeneous region with a predominantly Palestinian population into a primarily Jewish space. The very power of such maps lies in their potential to shape their users' cognitive maps. As a result, many Israelis are unaware of the other territorial identities underlying the places in which they live, viewing alternative claims as false or as part of political propaganda. Similarly, international tourists in search of religious sites with such maps in hand will be steeped in images of a Judeo-Christian territory that contrast starkly with the Palestinian cities, sites, and topographies that dot the landscape should they raise their eyes above the edge of their maps. Whether they will imprint such cartographic visions onto the landscape, or recognize the fissure and tensions between realities on the ground and imaginary geographies on maps, will always lie in the eyes of the beholder.

Jewish Land

Besides maps of the Land of the Bible, the Yesha Council also produced and distributed a map entitled "Transportation Routes in Judea, Samaria and Gaza" which could be purchased in bookstores across Israel from 2001 until 2012 and was aimed at Israeli drivers, tourists, and settlers (see Figures 5.6a and 5.6b). The map includes authoritative tourist maps for Judea, Samaria, and Gaza (the oPt). It was published by Carta, Israel's leading cartographic firm, which also supplies mapping services to commercial, academic, and governmental institutions.

(a)

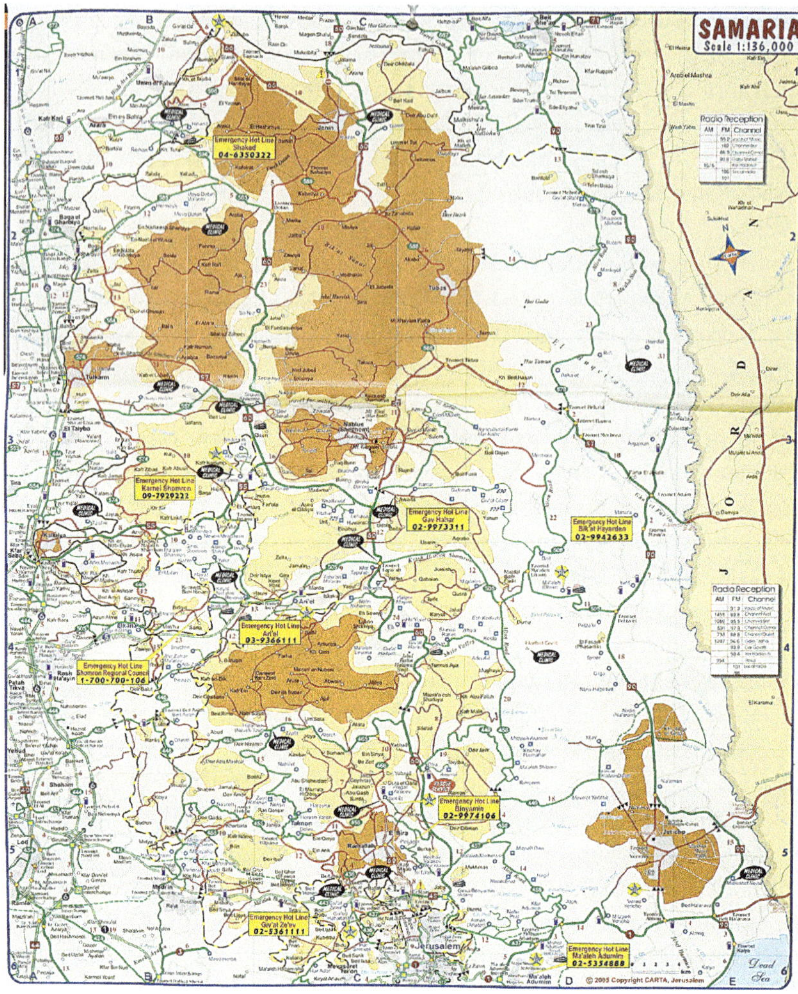

Figure 5.6a and 5.6b. Carta Map of "Samaria Scale 1:136,000" and Map of "Judea Scale 1:156,000"—part of a tourist map entitled "Judea, Samaria and the Gaza Strip: Major Routes, Settlements and Civilian Outposts." © Copyright Carta, Jerusalem.

Because drivers used this map to navigate the West Bank, cartographic details are extensive and include the internationally recognized spatial divisions of Zones A, B, and C as set out in the 1993 Oslo Accords. For Israeli drivers it was crucial to be aware of these divisions, as under Israeli law they are not allowed to enter the Palestinian-controlled Zone A.

Yet even within those seemingly more scientific maps, various visual and textual signifiers indicated the Yesha Council's ideological commitments. In

(b)

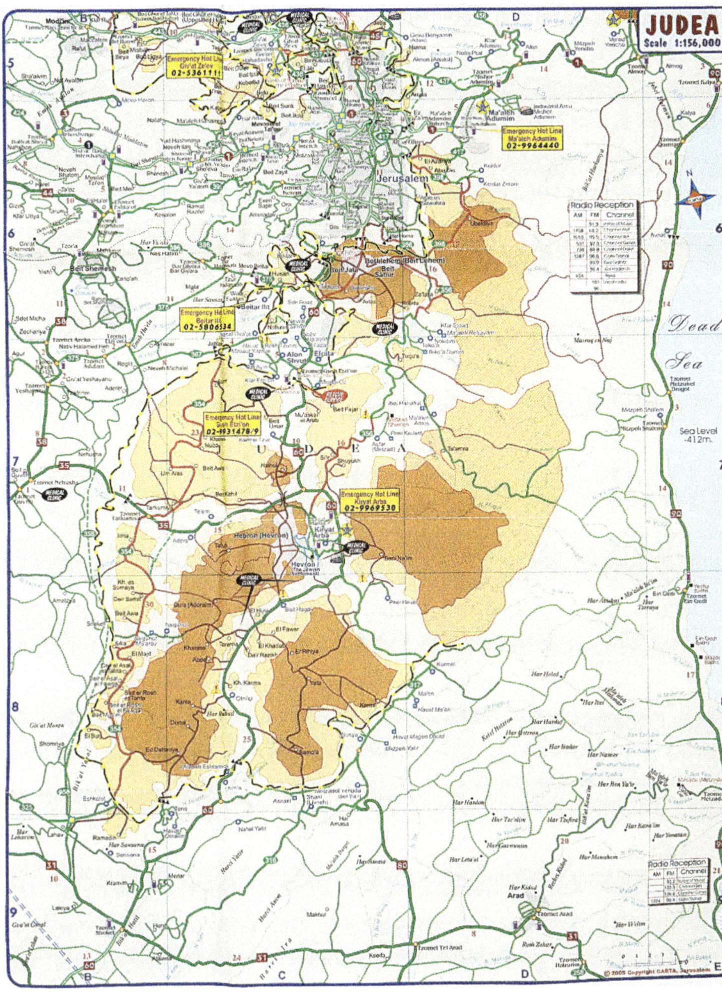

Figure 5.6a and 5.6b. Continued

the map, the West Bank and Gaza are represented as a direct continuation of the state of Israel. This is achieved by three complementary means. First, the map, in line with official Israeli maps (for example, the *Atlas of Israel*), omits the Green Line. Instead, in the heading, the West Bank is called Judea and Samaria, whereby it is presented as part of the Israeli national space. Second, the map eliminates the separate territorial status of the Palestinian territories by presenting the road system as contiguous with Israel's national road network. Third, the same set of colors, symbols, and level of details extend over the Israeli and the Palestinian territories. For instance, Israeli and Palestinian towns and villages, irrespective of their location to the west or east of the Green Line, are marked with the same symbol—a black circle, yet Jewish settlements and outposts appear as blue circles or squares with their names marked in blue thick letters. As "blue" signifies Israel's blue-on-white national flag, the choice of color serves to support the settlements and integrates them into the Israeli national space (despite the fact that according to the Fourth Geneva Convention any transfer of civilians for settling in occupied territories is illegal). Also, the complementary colors denoting Zones A, B, and C (with the same beige color extending from Israel to include Zone C) signify the perceived unity of the whole territory. Consequently, the level of cartographic detail (including color, symbols, and road networks) unifies the national spaces, while the heading claims the territories as part of Israel.

As this map was first published during the second Intifada in 2001, it organizes the West Bank's spatial order according to the perceived safety or dangers to the maps' target audience: Israeli drivers and settlers. As a result, the map hierarchically divides the territory into risky and safe spaces. Sizable yellow rectangles appear mainly near Zones A and B, providing an "Emergency Hot Line number" for the nearby areas. Zones A and B are much more densely equipped with emergency services than the Israeli-controlled Zone C, which is considered a "safe" zone. The map informs drivers of the location of emergency centers, police stations, rescue and medical services, and military roadblocks. Medical clinics and rescue services are indicated by large letters enclosed by black or red circles. These signs and symbols stand out as some of the most visible elements in the map.

Roads are further differentiated in terms of their safety for Israeli drivers. They are distinguished by color—green roads are considered safe for Israelis, as Palestinian drivers were prohibited or restricted from using them. In the map's key, they are described as "Road open to regular traffic (always check in advance)." Red roads, on the other hand, are marked as "No entry—danger ahead." These are roads used by Palestinians and therefore are assumed to signify potential dangers to the map's target audience. These color codes represent Israel's "Forbidden Road System." Accordingly, there are three types of "green" roads within the West Bank: roads on which Palestinian travel is completely prohibited, partially prohibited, or restricted (B'Tselem 2004). The green roads frequently bypass Palestinian towns and either do not have, or have blocked off, exits to Palestinian towns. Red roads, on the other hand, enter Palestinian-controlled territories, bypassing Jewish settlements. The road system spatially organizes the West Bank, channeling Palestinians and Israelis to use separate road systems.

The use of such visual symbols creates the perception of the oPt as the wild frontier of the Israeli national space (Schnell and Mishal 2008), while various verbal and textual signifiers construct the Jewish settlers as pioneers, and Palestinians appear as a threat and danger to Israelis' safety and security. Indeed, in the English-language version of the map, a note to the traveler lays out "emergency regulations for civilian travel in Judea, Samaria and the Jordan Valley." These regulations are said to apply to Israeli citizens, and all foreign visitors are advised to follow the same rules. Recommendations suggest avoiding travel on any road not marked on the map, except for streets in Jewish settlements. Travelers are advised to "1. Use vehicles equipped with stone-resistant windshields and a 'Barak' distress signal system. 2. Travel with at least one other vehicle. 3. Carry a personal firearm. 4. Carry a means of communication (at the least, a mobile phone)."[9] Also, while entering Zone A is "strictly prohibited," Zone B can only be crossed via the major Israeli bypass roads. Unmarked roads, and the paved or dirt roads that travelers are not supposed to enter, thus signify the uncomfortable silences in the map—an alternative topography and a people that have been erased from its markings, yet nevertheless are seemingly lurking in the maps' fissures ready to jump on unsuspecting tourists and Israeli drivers.

The ideological commitment of the maps' producers is clearly displayed in the Hebrew version of the map (as shown in Figure 5.6c).

On its cover, settlements of white red-roofed houses near water towers adorn an empty and expansive, hilly green and beige landscape, while the horizon is lined with a green forest. The red-roofed houses, the water towers, and the planted forest symbolize the allegedly pioneering spirit of early Zionist settlers and their ability to make "the desert bloom." As discussed in chapter 2, Zionists continue to maintain that they had returned to "the land of their ancestors—[which was] a sparsely populated desert wasteland described in the 1860s by Mark Twain as "a desolate country . . . given over wholly to weeds—a silent mournful expanse."[10] The "desolate wasteland," however, had been turned into fertile land by the Jewish pioneers, and the purportedly empty and virgin territory has erased any signs of a Palestinian presence, now or in the past.

Yet, as Efrat Ben-Ze'ev (Ben-Ze'ev 2011) points out, the perceived fertility of the land may have had less to do with the land's actual cultivation than with the elements cartographers had chosen to indicate on British Mandate maps. While they marked agricultural land in green, other cultivations such as local tree plantings or seasonal plantations were not marked as fertile green land, leaving the impression of large expanses of empty, uncultivated, and desert-like terrain,

[9] "Judea, Samaria & The Gaza Strip: Major Routes, Settlements & Civilian Outposts" (Carta n.d.) (Source: Carta).

[10] AIPAC: The American Israel Public Affairs Committee "Quest for Statehood," available at http://www.aipac.org/en/why-israel-matters/quest-for-statehood. Retrieved October 7, 2012.

Figure 5.6c. Front Page of Hebrew Version of the Carta map: "Judea, Samaria and the Gaza Strip: Major Routes, Settlements and Civilian Outposts." © Copyright Carta, Jerusalem.

supporting the orientalist assumption that Arab-Palestinians led the land fall fallow (see Davis 2011). The perceived infertility of a lot of the land may thus have had more to with map-makers' use of color, symbols, and cartographic choices than with the land as experienced and cultivated by Arab-Palestinian peasants.

Such map-making practices followed in the footsteps of 19th-century European explorers in search of religious and symbolic features significant to Western civilizations. They also had an "amazing ability to discover the land without discovering the people," (Doumani 1999, 15) but instead "discovering" a supposedly empty landscape inhabited by a few "natives" that were seemingly "remnants of a passing traditional society" (Doumani 1999, 14). It is indeed the cartographic distinction between green cultivated lands and the expanse of desert—the notion of an empty and neglected land—that Zionists adopted; subsequently they hailed it as their own achievement to have "made the desert bloom." Yet, the way maps formed Zionists' imaginings of the land, as well as how the Europeans' preoccupation with Christian religious sites was often at the expense of comprehending "the other," are long forgotten. Instead, by drawing on powerful Zionist symbols and myths, the settler's council legitimizes settlement expansion. This message is underscored by the head of the Yesha Council, Adi Mintz, in an accompanying statement on the map brochure:

> Despite the period of war in the land (Israel) in general and in Yesha (Judea, Samaria and Gaza) in particular, Yesha settlements are growing and flourishing. Likewise, road traffic throughout Yesha is flowing smoothly and continuously. We present you with a map of Yesha's transportation routes, for safe and quiet/peaceful driving throughout our land spaces (see Figure 5.6c).

The statement points to the success of the settlement project, stresses the smooth flow of movement in the area (yet without consideration for the consequent restrictions to the freedom of movement for Palestinians), and emphasizes that the area is part of "our land spaces." Additionally, an advertisement for the settlers' housing office declares: "Yesha, a victorious settlement project; 17,000 settlers have joined the settlements last year."[11] A range of visual and textual signifiers thus reconfigures the West Bank as a primarily Jewish space, reflecting the creation of a new spatial reality on the ground as well as within its cartographic renderings.

By 2012, the tourist map for Judea and Samaria and the Gaza Strip was no longer available. In 2009, Carta instead produced a map titles *Israel (and Autonomous Areas) Physical Map* which included what (on its cover) were called "Autonomous Areas."[12] It was the most detailed map of these areas available. Yet the detail was minimal. While during the threat of the second Intifada the maps accounted for the presence of what were perceived to be hostile elements within Judea and Samaria and Gaza, now these elements had seemingly been domesticated and moved into small shaded delineated areas indicated in the legend as "Palestinian autonomous areas." They faintly decorate

[11] Translation from Hebrew was done by Izhak Schnell and Ilil Naveh-Benjamin.
[12] Carta 2009 Israel (and Autonomous Areas) Physical Map 1:270,000 (Jerusalem).

the background of the physical map that is characterized by a clearly demarcated road network that connects the Jordan Valley to the Israeli coast. Some Palestinian cities survived their cartographic eradication as they are marked by name, such as Ramallah, and are indicated in the legend as a built-up area. The Palestinian name for the city of Nablus, while marked, is overshadowed by the Hebrew name for the city: Shechem. It seems as if the threat of the Palestinians had passed and their invisibility had become enshrined in the maps. For those traveling in the area, the map-makers also drew a thin yellow line with black markings symbolizing the Separation Wall (the lines were either continuous or intermittent depending on whether the wall was already built or planned). Yet the black-on-yellow lines do not appear in the legend. While this omission means the map-makers cannot be faulted for not representing what is on the ground, omitting an explanation of the symbols' meaning nevertheless has political implications. Only those who have detailed knowledge of the area and of the geopolitics of the region would even know what the faint yellow line with black markings stands for.

A 2010 *Carta Atlas* also "Israelized" the land to an unprecedented degree. Unlike some other maps, the *Atlas* does mark Zones A and B and indicates the "security fence," which also appears in the legend. Alongside the many Israeli attractions mentioned that define the Hebraized and "Israelized" territory, the city of Ramallah is one of the few Palestinian features to be marked. It is described as an "Arab city: government and administrative center of Palestinian National Authority (PNA) (also known as the Palestinian Authority (PA)). Some claim it to be site of biblical Ramah" (Carta 1980, 20). It is thus Ramallah's possible connection to biblical Israel that makes the city worth mentioning among the relevant places and sites of Israel. Unlike the physical map, this atlas does superimpose on the maps' sections of Judea and Samaria a warning saying "Because of changing circumstances kindly check with the authorities before entering Judea and Samaria." So, although the perceived risk and danger of the Palestinians to Israeli citizens are intermittently indicated, they are nevertheless increasingly written out of Israeli maps. Indeed, the Palestinians and their cities and their topography have become more or less completely invisible among the roads, sites, and cities represented.

Such tourist maps of roads, sites, and topographies, when produced by commercial Israeli companies, tell particular stories about ways to see and experience Israel. Indeed, Israeli-produced maps tend to reproduce a vision of Israel within its post-1967 boundaries with the Green Line erased and with the West Bank, Gaza Strip, and Golan Heights firmly located within the Israeli spatial imagination. The maps generically Hebraize the region, emphasize its Judea-Christian heritage, marginalize or eliminate any Palestinian traces from the map, and minimize the demarcation of Muslim attractions. Yet, at times, they include patches of color where Zones A are located so as to warn map-readers from entering such areas. Their marked tourist attractions generally correspond to what are considered to be "secure" areas in the West Bank with "less secure" areas devoid of such attractions. On the other hand, Holy Land maps tend to eliminate any geopolitical issues and instead speak only to the region's rich biblical history (Collins-Kreiner and Mansfeld 2005; Collins-Kreiner, Mansfeld, and Kliot 2006). These maps are devoid of national boundaries and depict Judeo-Christian sites within present-day Israel, Jordan, the Palestinian Territories, Syria, and Lebanon. They are most often highly abstracted delineations of the territory, adorned with images of tourist sites, and devoid of cartographic measurements of spatial relations between various sites. Consequently, places are seemingly detached from local topographies and geopolitics; instead these maps construct a religious reality that transcends the realities on the ground. These two generic forms of maps, the tourist map and the Holy Land map, both tend to be reproduced by Israeli cartographic services that produce maps for the purpose of touring the region.

Besides producing its own maps, the Yesha Council also collaborates with the internet start-up MyIsrael. This organization is run by online pro-Zionist activists who use social media, including Twitter, Facebook, and YouTube videos, to put forth pro-Israel, pro-settlement, and pro-Zionist messages, and wants to spread "the voice of reason and truth about Israel." They introduce themselves as a group for whom,

> the internet is the new public square where everything is being talked about.... We can't ignore the potential power of the internet in helping us spread Zionist pro-Israel information around the world so that everyone can know that the Land of Israel belongs to the Jewish people.[13]

MyIsrael exemplifies the mushrooming of political interest groups that use social media to spread their geopolitical visions and disseminate maps to claim national territories for the Jewish people. Their alleged neutrality, willingness to combat "misinformation," and aim to promote "reason and truth" provide them with powerful tools to persuade a public that is not steeped in the political complexities of the pro-Israel debates. Indeed, just as MyIsrael promotes its pro-Israel image, so do other organizations that propagate territorial compromise with the Palestinians, such as Peace Now, B'Tselem, and J Street.[14] Yet the

[13] See MyIsrael, http://www.myisrael.org.il/action/en/about. Retrieved October 5, 2012.
[14] For J Street's website, see http://jstreet.org/.

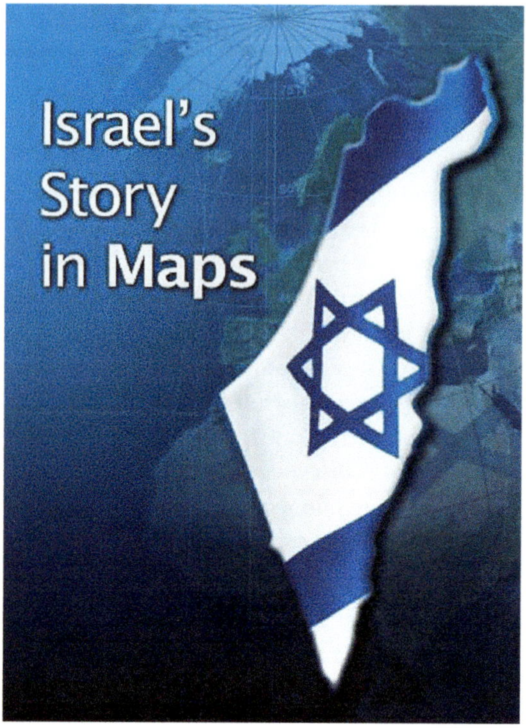

Figure 5.7. My Israel's Logo Map. Available at http://www.myisrael.org.il/action/en/.

political assumptions underlying their respective pro-Israel proclamations are fundamentally opposed. Their differences come down to maps. When MyIsrael advocates argue for the simplicity, accessibility, truthfulness, and representativeness of maps, maps become particularly powerful tools to put forth their ideological vision. For instance, MyIsrael's online publication of an atlas appeals to a right-wing, neo-Zionist conception of the Land of Israel. Its cover page consists of a "logo" map of a Greater Israel that is covered with the Israeli flag, as shown in Figure 5.7.

Such widely distributed logo maps, which are disseminated with the allure of truthfulness yet communicate a particular right-wing ideological position, can powerfully shape a public's imagination and conception of territorial realities. Indeed, the rhetorical trope of speaking in the name of science, objectivity, and truthfulness is still one of the most effective ways for various organizations—whether from the right or the left—to put forth certain politically inflected views. A relief map within the *Atlas* also visualizes their political stance and framing of their narrative, as revealed in Figure 5.8.

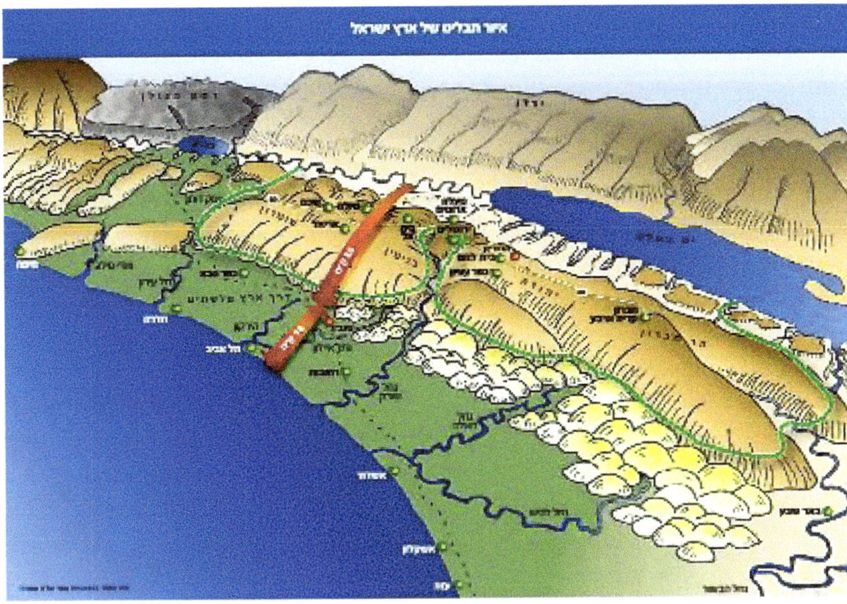

Figure 5.8. A relief map of Israel.
Source: MyIsrael. Available at http://www.myisrael.org.il/action/en/.

The simplified topographical maps show the green and fertile plain of the lower basin and the mountainous terrain of Judea and Samaria. The map is textually framed so viewers can understand the message communicated through the maps' vivid colors and symbols. The text points out that only 3% of Judea and Samaria is built up, yet it is rich with Jewish historical sites, surrounds historic Jerusalem from all three sides, and is populated with over 330,000 Jews (one-third ultra-orthodox, one-third national religious, and one-third secular). Geographically, Judea and Samaria are presented as the water reservoir of Israel; they are a central crest of land (55 km wide) that is surrounded by valleys, and that is much larger than the width of Israel's coastal plain which, from the edge of the Mediterranean to the Green Line constitutes up to 16 km. The Cross-Israel Highway Route 60 that allegedly follows the main biblical road along the mountain crest, crosses the West Bank, connecting Israeli cities in the south, such as Beer Sheba, to cities in the north, like Megido. There are only a few references to a Palestinian presence in the West Bank, and these emphasize that 43% of the territory constitutes Zones A and B, containing 90% of all Palestinians, and it is democratically governed by the PA.

Such narratives exemplify the rhetorical tropes used by the settlers' lobby in regard to Judea and Samaria; they emphasize Jewish historical roots in these territories and the land's abandonment by the Palestinians. Their emphasis on the heterogeneous nature of the settlers (ranging from orthodox to secular) underscores their united and collective effort to settle the biblical territories, which are also seen as strategically important to Israel's security. This rhetoric hereby brings together the "secular expansionist" argument for securing Israel with "national religious" sentiments for redeeming the land, which also informs the narratives of the new rightist coalition of the Likud and of religious parties.

As we have seen, Likud's and MyIsrael's political propaganda maps as well as the Yesha Council's tourist maps tend to all adhere to commonly shared visual and textual narratives. Whether the maps are abstract or cartographically detailed, by highlighting or omitting certain elements, they either communicate concerns for Israel's national security or a belief in Jewish rootedness in a biblical homeland. The very selectivity of the maps and their narratives hereby inevitably turn them into politics by other means. Yet, besides political parties and organizations that express commonly held right-wing visions of the territory and of the Israeli-Palestinian conflict, there are also various human rights and peace-promoting organizations (that constitute what is known as the "peace camp") that also create maps. It is to these organizations from the Israeli left, such as Peace Now, B'Tselem, and the Geneva Initiative that we now turn.

6

Maps with a Mission Post-1967

Israeli "Peace Camp" NGOs on the Warpath

Introduction: Maps in the Name of Peace

With the pre-1967 relative consensus in regard to Israel's boundaries fractured after the 1967 war, it was not only right-wing interest groups and organizations that increasingly used maps to make territorial and geopolitical claims; a number of left-wing peace and human rights organizations, here referred to as the "peace camp," increasingly started to do so, too. Maps produced by the "peace camp" are informed by a range of very different discourses, which include concerns about Israel's occupation strategies, its compliance or non-compliance with international law, its demography, the need for the recognition of Palestinians' human rights and historical presence in the region, and the feasibility of particular territorial solutions. Most organizations use a mixture of textual and visual signifiers as well as spatial demarcations that signify a map's adherence to international law and scientific cartographic conventions.

Such organizations frequently use maps produced by the United Nations Office for the Coordination of Humanitarian Affairs (UN OCHA) in Jerusalem as base maps. UN OCHA produces an array of very detailed and regularly updated maps of Israel and the Palestinian Territories.[1] UN OCHA maps tend to carefully reflect international treatises and laws, ranging from the 1949 armistice line to the Oslo Interim Agreements. UN OCHA's mandate is to be strictly neutral, impartial, and non-political and to share its data with Israeli as well as Palestinian organizations. For "left-wing" Israeli NGOs, such as Peace Now or B'Tselem, the very assumption of the neutrality, impartiality, and universalism of UN OCHA-produced maps provides their political struggle with legitimacy.

Their cartographic precision and adherence to international law is exemplified in UN OCHA's closure maps. These maps delineate the territory according to international law by drawing the 1949 armistice line—the Green Line—as the internationally agreed upon division of Israeli and Palestinian Territories. They also demark territory in accord with the Oslo Interim Agreements that divide the territory into Zones A, B, and C. They further indicate Israeli settlements

[1] For additional maps see UN OCHA http://www.ochaopt.org/maps. Retrieved January 25, 2020.

The Politics of Maps. Christine Leuenberger and Izhak Schnell, Oxford University Press (2020).

DOI: 10.1093/oso/9780190076238.001.0001

(that contravene international law) that are beyond the Green Line, and they mark the West Bank Barrier (the "Barrier" is considered a more neutral term that terms used by protagonists, such as "fence" or "wall"),[2] which deviates from that Green Line (see Figure 6.1). Such maps also distinguish various Israeli closure mechanisms inside the West Bank and trace the separate Israeli- and Palestinian-controlled road systems. In fact, the roads are visually severed by the prominence of the Green Line and the barrier, making clear that the road infrastructure, rather than integrating the two territorial spaces, serves to differentiate them from one another.

Different cartographic layers of such maps are often used by NGOs to provide their own maps with the cartographic and legal clout that they associate with the UN OCHA maps.[3] We now turn to some of the Israeli NGOs that tend to align themselves with the international consensus on territorial delineations and rely on international law as a persuasive tool in their political struggle.

One of the most prominent groups within the peace camp has historically been the Israeli organization Peace Now. This NGO was founded in the late 1970s and supports governmental and non-governmental efforts to seek reconciliation with the Palestinians through exchanging "land for peace." Peace Now opposes and monitors settlements within the oPts, and attempts to raise public support for the peace process. Another leading human rights organization, B'Tselem (established in 1989 in Israel and in 2008 in the United States), aims to educate the Israeli public about human rights violations in the oPts, urges Israel's compliance with international law, and attempts to influence Israeli and American policymakers. The Geneva Initiative[4] is an organization that came into being in 2003, 10 years after the Oslo Accords had been agreed to but still not implemented. They present a model for a permanent status agreement between Israel and a future state of Palestine and propose solutions to a number of contentious issues in order to end the historic conflict and realize a two-state solution. Last, Zochrot is one of a

[2] International media outlets, such as the BBC as well as the UN, use the terms "separation barrier" or "barrier" as acceptable generic terms that avoid the political connotations of "security fence" (preferred by the Israeli government) or "apartheid wall" (preferred by the Palestinians). BBC. Available at https://bbcwatch.org/2013/03/26/the-politics-of-bbc-approved-terminology-on-israels-security-fence/. Retrieved May 31, 2019. See also Leuenberger 2016b.

[3] This is not to say that various Israeli and Palestinian protagonists don't find fault with UN OCHA maps. Indeed, both Palestinian and Israeli interviewees have criticized these maps as top-down maps that often show a failure to understand local realities. For some Israeli cartographers interviewed, the UN is not a neutral arbitrator and their depiction of the Green Line is considered to be political and not in line with facts on the ground, primarily the fact that the Green Line ostensibly lost its relevancy after the peace agreement with Jordan. On the other hand, Palestinian cartographers often disagree with some of the information, terminology, and cartographic detail included in these maps, such as how to refer to "the Separation Wall," how to indicate its territorial effects, and how to refer to Jewish settlements in the West Bank.

[4] See the Geneva Initiative http://www.geneva-accord.org.

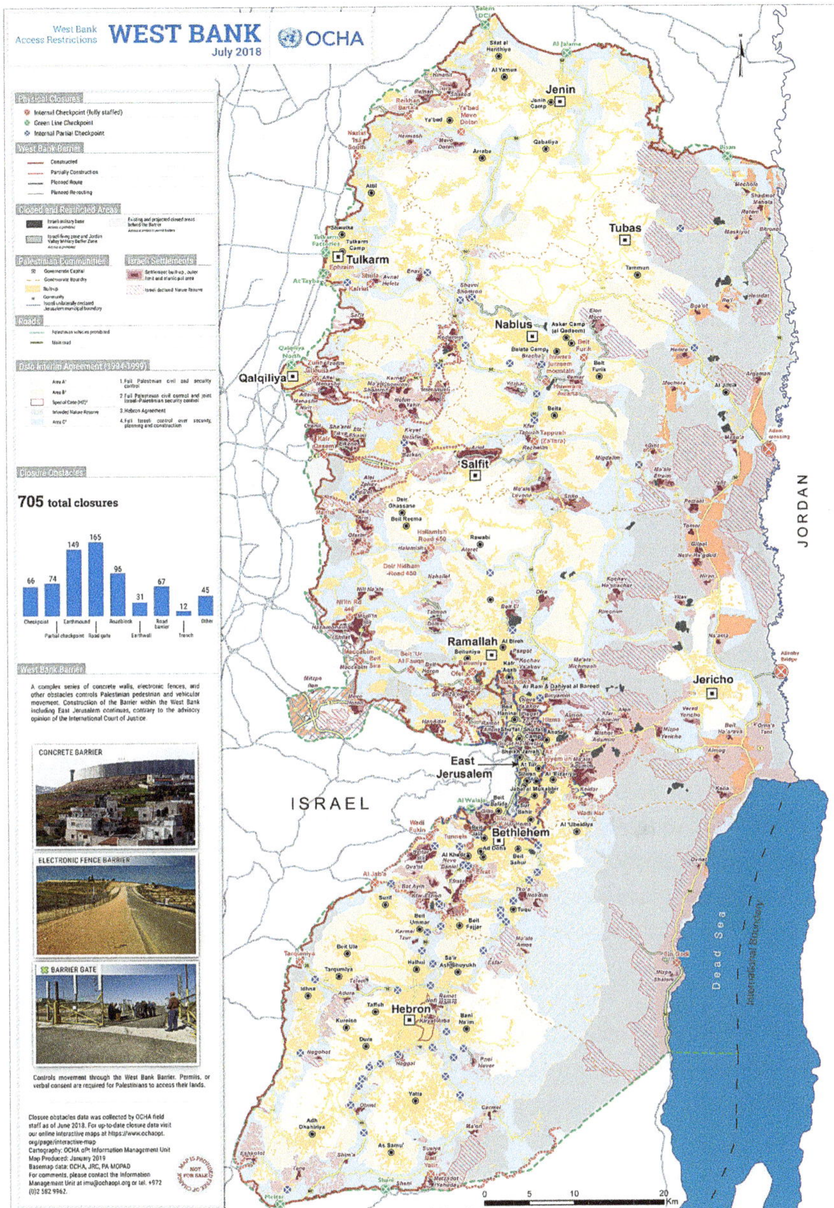

Figure 6.1. UN OCHA West Bank Access Restrictions, July 2018. Available from UN OCHA, https://www.ochaopt.org/maps. Retrieved May 31, 2019.

number of "bottom-up" mapping initiatives[5] that challenge the dominant Hebrew topography in maps by mapping an alternative Arab topography. All these organizations regularly produce maps in order to raise awareness about certain social concerns and to further particular political aims.

Peace Now

Peace Now produces various maps that emphasize the territorial strategy of the Israeli occupation and the demographic imbalance between Palestinians and Jewish settlers inside the oPt. Demographic arguments are used to highlight the need to "separate and disengage" from the Palestinians in order to preserve Israel as a Jewish and democratic state. While this ethnically nationalist rhetoric tends to be associated with Israel's right wing (and often sits uneasily with Peace Now advocates), Peace Now nevertheless uses demographic data to advocate a return to the Green Line, not least because demographic concerns resonate powerfully with Israeli public sentiments and with their vision of a democratic state for the Jewish people. An informational leaflet produced by Peace Now entitled "Divide and Rule: The Ongoing Israeli Rule over the West Bank"[6] contains a detailed map, "The Israeli Model of control over the West Bank" with information about Israel's strategies of control in the oPts and data on Israelis' attitude toward settlement activities.

In the map, depicted in Figure 6.2, the basic outline and name of the territory "the West Bank" is in line with maps produced by UN OCHA (UN-OCHA 2007) in terms of their adherence to legal treatises pertaining to the territories. The delineation of the boundaries is based on the international legal convention that territories cannot be annexed by force and borders can only be declared based upon peace deals. Thus, unlike the Yesha Council and other right-wing groupings, the territory is not "Israelized" by naming it Judea and Samaria, but rather, Peace Now maps represent the West Bank as separate from any other territory surrounding it. This is achieved by clearly demarking the Green Line and by differentiating the area from Israel and other neighboring countries by visual signifiers including color and level of cartographic detail shown (e.g., roads and cities). Also, unlike the Yesha Council tourist maps in Figures 5.6a and 5.6b, this map further emphasizes the West Bank's separate status by color-differentiating Israeli territory from the C Zone.

[5] Another NGO that does bottom-up mapping is Grassroots. For more information on this organization see http://www.grassrootsalquds.net/.

[6] The concept of "divide and rule" is associated with British colonial policies in the region. Its use embeds an explicit political meaning within the map.

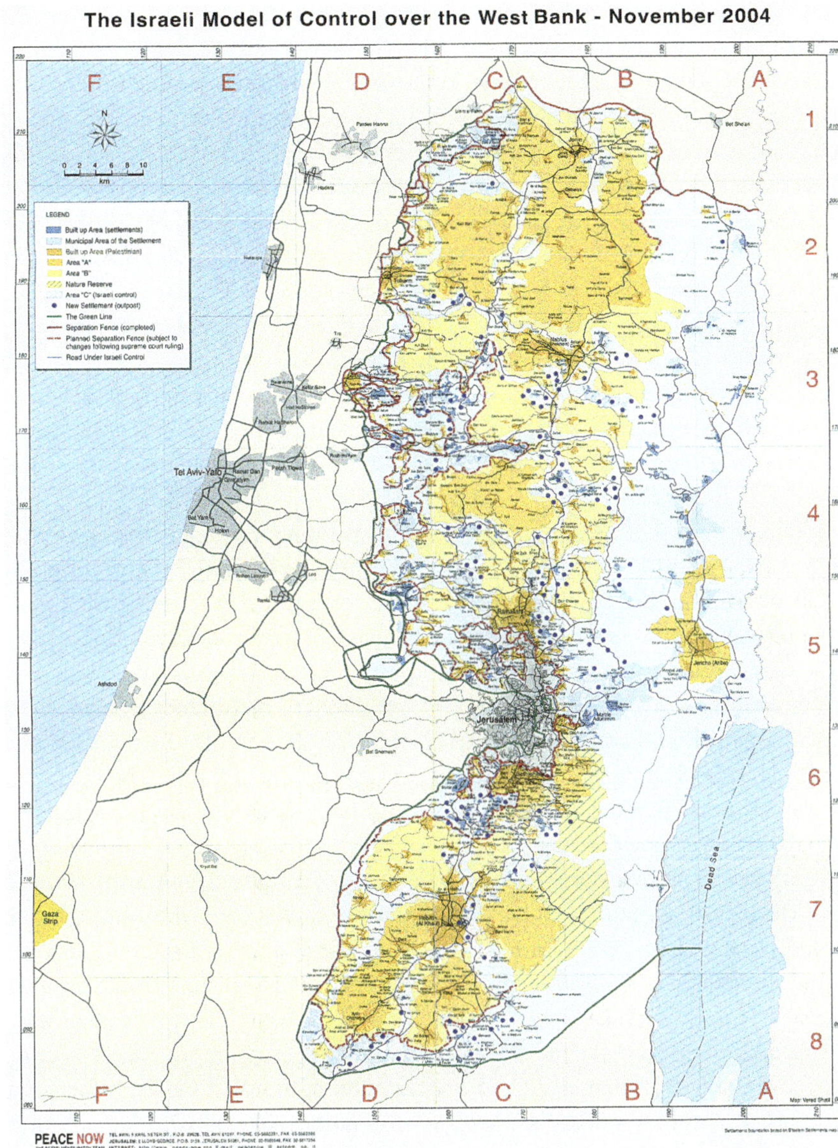

Figure 6.2. Peace Now's information leaflet "Divide & Rule: The Ongoing Israeli Rule over the West Bank" contains a map entitled "The Israeli Model of Control over the West Bank—November 2004." Available at http://peacenow.org.il/site/en/peace.asp?pi=59&fld=60. Retrieved March 5, 2010.
Source: Peace Now.

In adopting the internationally legitimized delineation of the territories and by recognizing the separate territorial status of the West Bank, Peace Now aligns itself with the current international consensus. Unlike other Israeli maps that represent the area around Latrun (in the lower half of the West Bank) as part of Israel, Peace Now maps tend to mark this area (in line with the UN maps) as "No-Mans-land."[7] Besides adhering to internationally established conventions for depicting the territory, the Peace Now map ("The Israeli Model of Control over the West Bank") also closely follows cartographic conventions. For instance, the names of towns and villages in both Palestinian- and Israeli-controlled areas are similarly marked, adhering to cartographic principles of size, type of lettering, and population density. The persuasiveness of the map is thus partly based on its appeal to scientific accuracy. This is apparent within the map's narratives, where the cartographers provide detailed information from public opinion polls.[8] Such evidence-based arguments are a powerful way to attempt to increase the map's legitimacy and credibility in the eyes of its readers.

While the map is constructed according to cartographic norms, the large blue circles that are dotted all over the map (signifying unauthorized Jewish outposts in the West Bank) deviate from cartographic convention in terms of their size. The circles, however, serve as a visually effective tool for politically critiquing the settlement projects. The level of cartographic detail within the West Bank also serves the map-makers' aims. By clearly marking the hierarchical road system with blue Israeli-controlled roads and black Palestinian roads, the map stresses the Israeli model of control over the territories (B'Tselem 2004). The map also indicates the different territorial statuses of Zones A, B, and C. While the fully or partially Palestinian-controlled Zones A and B are distinguished with complementary yellow colors, Zone C appears in blue to highlight Israeli control of the area. These color variations differ from the cartographic conventions of the UN OCHA map, which presents Zone C in white and as a "closed military area" to stress its occupied status and separateness from Israel. However, Peace Now uses the color blue to emphasize the expansionist policies of the Israeli government. In this map, then, the producers have used various cartographic tools to construct a credible argument against the occupation of the West Bank. By focusing on Israeli-controlled roads, the location of unauthorized Jewish outposts, and settlements, and by "Israelizing" Zone C with the color blue, the map points to how Israeli territorial control divides, splits, and fragments the West Bank into

[7] However, on the Peace Now website (www.peacenow.org.il) some Hebrew-only maps present the "No-man's-land" around Latrun as part of Israel. To what extent Hebrew-only maps differ from English-language maps, and how such differences link to the maps' target audiences remain to be investigated.

[8] The polls were conducted by academics associated with Peace Now, including one of the co-authors, Izhak Schnell.

separate Palestinian-controlled blocks that lack any territorial continuity. Indeed the patches of yellow disappear within a sea of blue.

On the back of the Peace Now informational leaflet (titled Divide and Rule: The Ongoing Israeli Rule over the West Bank) are smaller maps of various Israeli proposals about how to divide the West Bank, including the "The Sharon Plan—2004" proposed by former Israeli prime minister and military leader Ariel Sharon (as seen in Figure 6.3).

This very abstract map, in Figure 6.3, contains few cartographic details, but it contrasts the colors blue and yellow to show the proposed division of the territory. The visual similarity of the "Sharon Plan—2004" map with the current territorial control of the area shows the relative success of that plan. In 1973 Sharon famously said:

> We'll make a pastrami sandwich of them. We'll insert a strip of Jewish settlements in between the Palestinians, and then another strip of Jewish

Figure 6.3. Peace Now's information leaflet "Divide & Rule: The Ongoing Israeli Rule over the West Bank" contains a map entitled "The Sharon Plan—2004."
Source: Peace Now.

> settlements right across the West Bank, so that in 25 years' time, neither the United Nations nor the United States, nobody, will be able to tear it apart. (Shatz and Carey 2004, 1)

By juxtaposing the Sharon plan to the map depicting current Israeli mechanisms of control over the West Bank, the Peace Now leaflet draws upon their visual similarity to support the argument that Sharon's plan has been largely implemented. Israel has de facto annexed more than 50% of the Palestinian territories. Indeed, between the yellows and blues a pastrami sandwich has been made. Israeli settlements, bypass roads, and checkpoints have truncated the territory to "the point of no return" (Halper 2004).

The map's accompanying texts aim to increase public awareness about settlement projects. A section entitled "Did You Know That?" points out that "since the signing of the Oslo accord over 100 new outposts have been established illegally in a continuing attempt to prevent any possibility that Israel and the Palestinians might reach a political agreement." The leaflet argues that the settlers within the West Bank comprise less than 3% of the Israeli population, while Israel invests 50 billion New Israeli shekels (NIS) in the settlements. It also notes that there is a lack of public support of the settlement project: "Only 11 percent of the Israeli public is convinced that new settlements should be established and 87 percent support the dismantling of the outpost," and even most settlers would have accepted dismantling the settlements in exchange for peace at that time (the willingness to stop the settlement projects and exchange land for peace has diminished substantially since 2004). The text further emphasizes that other plans for territorial division, such as the Geneva Initiative, would provide a more equitable distribution of land and "put an end to Palestinian claims to Israeli territory." In this informational leaflet, the visual signifiers that compose its cartographic details, together with the adjacent textual narrative, provide a coherent political interpretation of the maps.

Along with producing maps that emphasize Israel's occupation strategies in the West Bank, Peace Now also publishes demographic maps including a "Population Map of the West Bank and the Gaza Strip" as seen in Figure 6.4.

This map follows the same conventions as other Peace Now maps. It emphasizes the Palestinian Territories' discrete legal standing by visually emphasizing the Green Line. Also Zone C is marked in white rather than blue (signifying the blue of the Israeli flag), which again reinforces the West Bank's separate territorial status. The map's main concern is the demographic imbalance between the Jewish and Palestinian populations within the West Bank and Gaza. Different-sized blue and red circles signify the size of the Palestinian and Israeli populations in the cities and villages with the territories. The red circles that predominate portray the Palestinian population, and the small, scattered blue circles that indicate the Jewish population visually illustrate that small isolated

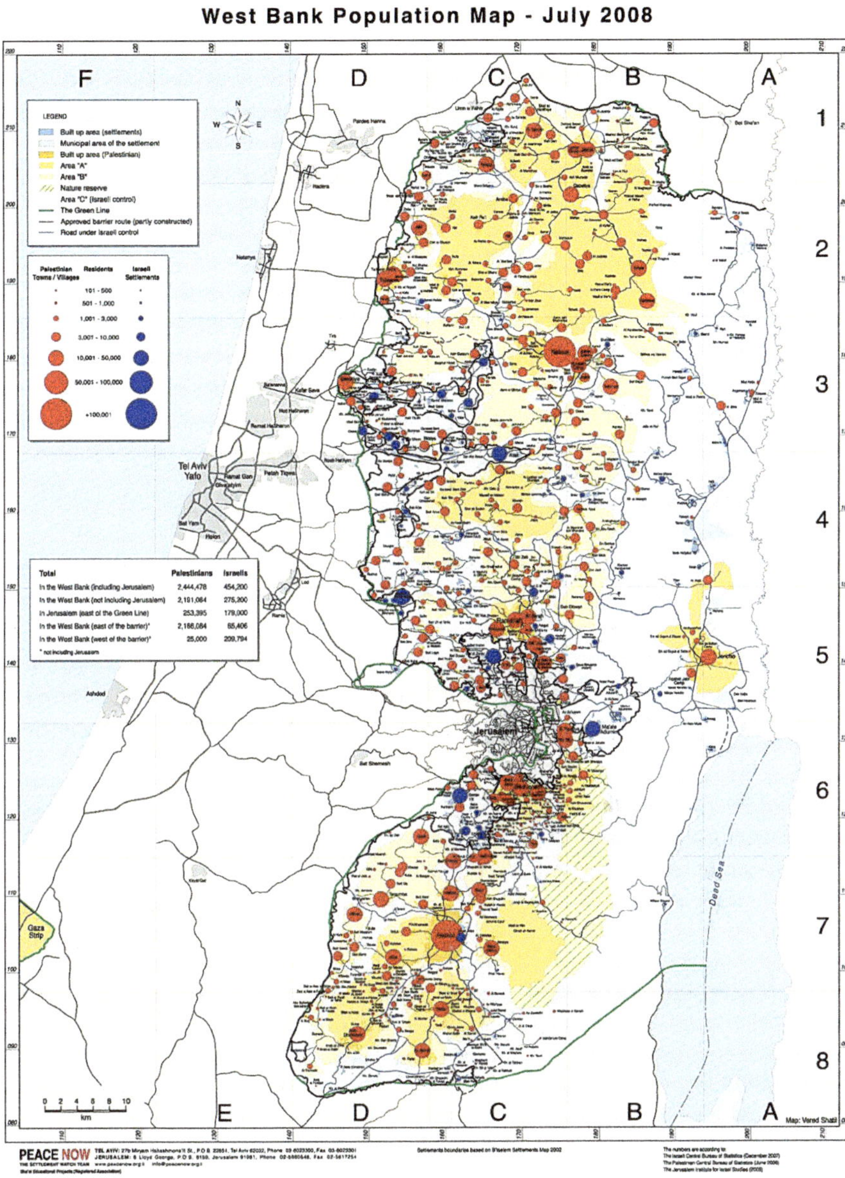

Figure 6.4. Peace Now: "Population Map," July 2008.
Source: Peace Now.

Jewish settlements are spread among highly populated Palestinian areas and point to the failure of settlers to create a Jewish majority in these areas. This map exemplifies the use of a demographic discourse by the political left as it argues for a disengagement from the Palestinian Territories. Similar kinds of demographic fear were one of the driving forces behind the disengagement from Gaza in 2005, particularly the concern, discussed in chapter 5, that if Israel was to remain a Jewish democratic state it couldn't absorb a much faster growing Palestinian population. In fact, Israel has a long history of attempting to contain what some refer to as the "demographic time bomb": Palestinians' higher fertility rates. And it is the threat of that "demographic time bomb" that is also a Zionist's nightmare.

Peace Now produces another genre of maps which can cartographically envision future scenarios and which may underscore the feasibility of certain courses of action. In 2002, in response to public debates as to how settlers in the territories would respond to the possibility of having to evacuate their settlements, cartographers at Peace Now mapped the results of an opinion poll with settlers on their attitudes toward withdrawal (as shown in Figure 6.5).

The maps were designed to help inform the public debate over the feasibility of evacuating settlers as part of a peace process. The left-hand map (as shown in Figure 6.5) represented settlers' "Expected Responses to Withdrawal Decision" and the right-hand map shows "Settler Readiness to Leave in Exchange for

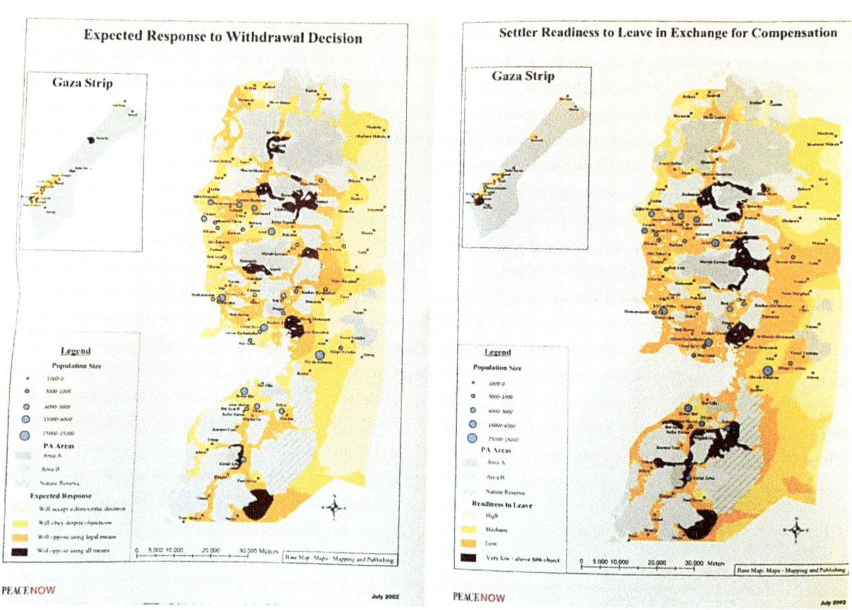

Figure 6.5. "Settler Attitudes towards Withdrawal from the Settlements," July 2002.
Source: Peace Now.

Compensation." The array of colors (ranging from yellow to dark brown) suggests that most settlers from the larger settlement blocks (whose primary motivation for settling in the oPts was economic) would accept the decision to withdraw and would leave the settlements if compensated. Only a small percentage of ideologically driven "national-religious" settlers (whose aim is to settle Greater Israel, resided in the dark brown area, and lived in close proximity to Palestinian residents) was expected to resist. These maps therefore suggested that apart from potentially some isolated violent incidences, evacuation of settlements as part of a peace deal with the Palestinians would be feasible. These maps generated a lot of publicity. Officials from the government and the military establishment wanted to be briefed on their findings. The maps' lead designer (co-author of this book Izhak Schnell) was invited to present its findings to veteran generals at the Israeli Defense Force and to the Office of the National Security Advisor, which was in the middle of planning for the withdrawal from Gaza. This map hereby demonstrates the political power of maps and how they can become part of political debates.

The Peace Now maps discussed here exemplify the use of various narrative and representational strategies to advocate a two-state solution. These kinds of discourses concerning Israel's occupation strategies and demography are frequently invoked as evidence by advocates for Israel's withdrawal from the oPts. Thus, the map-makers' integration of internationally agreed-upon territorial delineations and cartographic conventions undergird the legitimacy and credibility of the maps' messages. Other organizations within the "peace camp," including the human rights organization B'Tselem, uses similar mapping strategies to empower itself in a politically contentious environment.

B'Tselem

B'Tselem's mapping strategies are exemplified by a map entitled "The Separation Barrier in the West Bank February, 2008" (see Figure 6.6). The Barrier (also referred to as the "Security Fence," the "Segregation—[or "Separation"]—Wall," among other politically charged names) has been under construction by the Israeli government since 2002 and is one of the more controversial mega-engineering structures the government has undertaken (Leuenberger 2014). According to the government, the purpose of the barrier is to increase security within Israel, yet by its route it infringes on international law and it de facto annexes Palestinian land and water resources.[9] As a result, many Israeli NGOs

[9] For the Israeli government's rationale for the "Security Fence," see http://www.mfa.gov.il/mfa/foreignpolicy/terrorism/palestinian/pages/saving%20lives-%20israel-s%20security%20fence.aspx; for the International Court of Justice's Advisory Opinion on the "wall," see http://www.icj-cij.org/docket/index.php?p1=3&p2=4&case=131&p3=4.

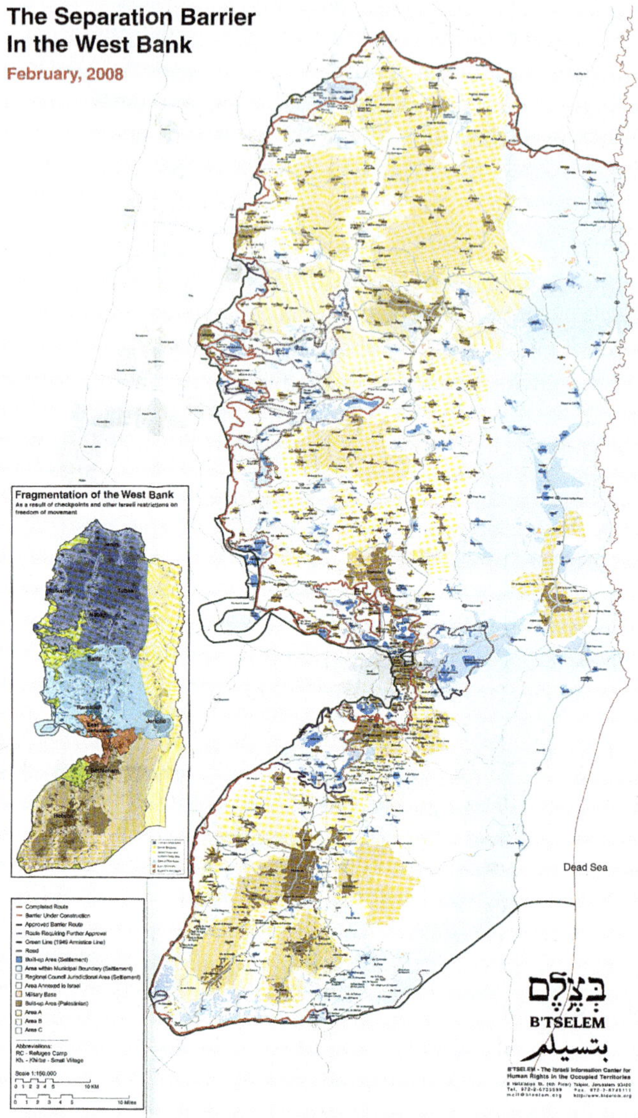

Figure 6.6. B'Tselem map: "The Separation Barrier in the West Bank," February 2008.
Source: B'Tselem.

trace its route to emphasize a particular political message, such as its deviation from the internationally recognized Green Line. B'Tselem exemplifies this practice. The map presented in Figure 6.6 visually demonstrates how Israeli policies of building the Separation Barrier, permitting settlements, enforcing the

"Forbidden Road Regime," and establishing Israeli military zones have split the territorial integrity of the West Bank. It also highlights the Israeli-controlled territorial corridors that connect settlements inside the West Bank to Israel, which further serve to split and divide Palestinian-controlled enclaves from each other.

This map also shows the Green Line and includes the delineation of the No-Man's land near Latrun, and, in doing so, reproduces internationally recognized cartographic conventions established in UN-produced maps. Its heading, "The Separation Barrier," also appeals to neutrality as the term Barrier carries less political connotations than other possible descriptors (Leuenberger 2016b).

Unlike more abstract logo maps, this map integrates cartographic details such as municipal and regional council boundaries. By doing so, B'Tselem can point to the relative size of the Jewish settlements in relation to the territory within their jurisdiction. As one interviewee, a B'Tselem officer, commented, "The dark blue is the built area, the medium blue are the municipal boundaries and the lightest blues are the regional councils of the settlements. So you see that the settlements in fact are relatively small. It is less than 2% of the West Bank, but the land controlled is more than 40%" (Interview M 2008). Patches of brown denoting Palestinian-controlled areas are submerged in a sea of blue, providing a visually effective means to stress the fragmentation of the West Bank and highlight the demographic imbalance between Jewish settlers and Palestinians in the territories. Integrating cartographic detail, such as municipal boundaries, can raise social issues important to the organization while enhancing the organization's claim to scientific credibility. B'Tselem has a reputation for reliability. Its researchers conduct extensive fieldwork and are known to regularly cross-check facts with numerous other organizations and governmental agencies. This, as one B'Tselem officer points out, is crucial: "We are a very vocal controversial organization . . . so it is very important that information is 150% accurate. Credibility is our calling card" (Interview M 2008).

By using cartographic conventions, the map's producers utilize scientific standards, instilling the map with authority and persuasive power. Visual and textual signifiers reproduce internationally recognized territorial demarcations. The map not only depicts the Green Line but also uses the words "West Bank" rather than Judea and Samaria and "settlements" (which contravene international law), and marks the "white" Zone C as "annexed to Israel" (thereby occupied). Like maps produced by the "secular expansionist" or the "national religious" camps, those produced by "peace camp" organizations, such as Peace Now and B'Tselem, use visual rhetoric. In order to retain credibility and prevail in a politically charged environment, however, the maps' producers attempt to transcend Israeli party politics and frame their social and political concerns by appealing to the science of cartography and by invoking international law. For the "peace camp," then, the appeal to international law and scientific cartography is an important aspect of their attempt to communicate impartiality and reach a wider public.

Geneva Initiative

Unlike maps that describe a current state of affairs or aim to communicate certain social concerns, outside threats, or possible future scenarios, maps have also been designed to serve as a basis for peace proposals. Indeed, the power of maps in negotiations is so compelling that during peace talks in July 2000, the Israeli negotiator Oded Eran was instructed by Ehud Barak not to show the Palestinians a map. He was only "authorized to show the Palestinians a schema. It wasn't a map, it was just a general outline" (Frontline 2002). It is also during negotiations with Israel that Palestinians found they were ill prepared, as they had "no maps of their own." Indeed the data and information they had "was scant when compared to what the Israeli negotiators had at their fingertips" (Scanteam and ARIJ 2009). Since that time Palestinians have developed their cartographic capacity so as to be better able to assess Israeli proposals concerning boundaries, land annexation, and territorial swaps as well as the sharing of Jerusalem. In other words, maps are the non plus ultra in territorial negotiations and essential for future peace negotiations. Indeed, historically, those who have the maps have always had the power (Day 2008). The Geneva Initiative is one of the few organizations that has enshrined its vision of a two-state solution cartographically. Its aim is to produce the maps necessary for a final status agreement that will end this long-standing conflict. Its map entitled "Israel-Palestine Permanent Borders" delineates the boundaries between a proposed state of Palestine and the state of Israel (as shown in Figure 6.7).

The map resulted from negotiations among Palestinian and Israeli leaders who supported the idea of reaching a peace agreement, a group that included negotiators who had taken part in the failed Camp David Summit in July 2000 between the President Bill Clinton, Israeli Prime Minister Ehud Barak, and the PLO's leader at the time, Yasser Arafat. Its proponents worked to produce a map that showed that a territorial solution was indeed within reach, given some territorial exchanges and compromises that would not substantially alter the allocation of land between the two sides. The map incorporates a proposal for Israel to annex high-density Jewish settlements within the West Bank within its borders in exchange for other, currently Israeli-held territories that would be transferred to Palestine. Indeed the map's use of color shows that, based on the principle of exchanging equal territorial units, only two narrow land strips (marked in blue and green, respectively), one on each side of the Green Line, would have to be exchanged so as to include up to 80% of settlers within Israel, while compensating Palestinians with other unpopulated Israeli territories. The No-Man's Land around Latrun (in the middle of the West Bank) would also be divided equally between the two sides (Arieli 2011). With these concerns accounted for, the map delineates a possible boundary that

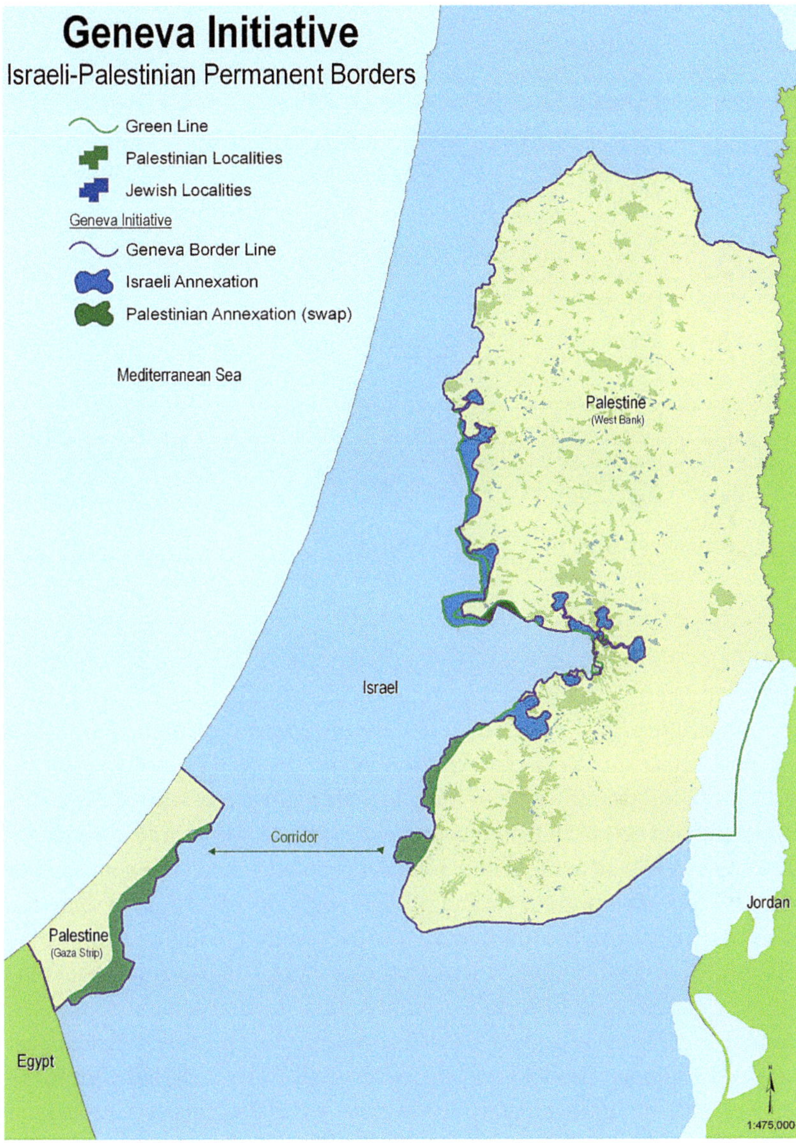

Figure 6.7. Geneva Initiative: Israel-Palestine Permanent Borders.

Source: Geneva Initiative. Available at http://www.geneva-accord.org/mainmenu/geneva-initiativeisrael-palestine-permanent-borders. Retrieved May 31, 2019.

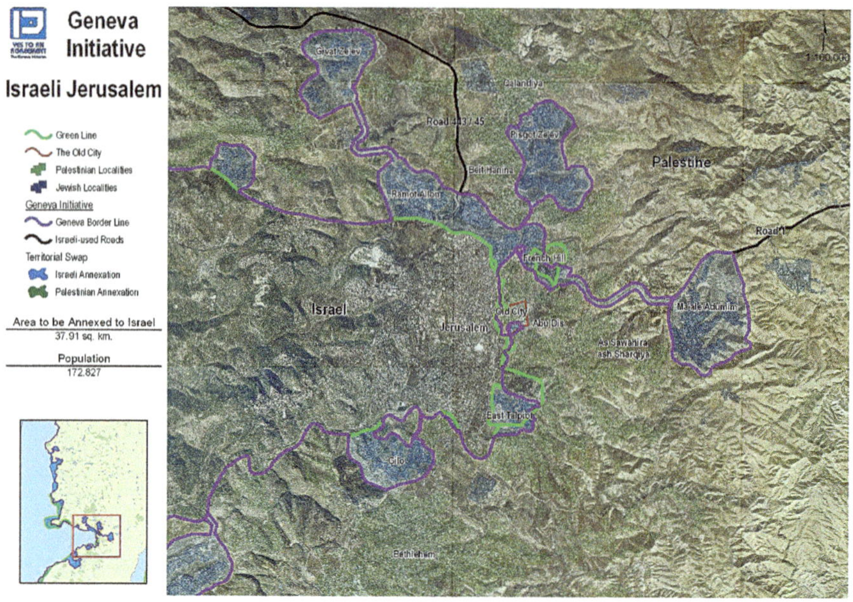

Figure 6.8. Geneva Initiative: Israeli Jerusalem.
Source: Geneva Initiative. Available at http://www.geneva-accord.org/mainmenu/geneva-initiative-jerusalem. Retrieved May 31, 2019.

could serve as an acceptable solution to the territorial conflict. While the map does not include a detailed discussion of the proposed agreement, its visual language nevertheless makes a powerful statement: the secret to solving this long-standing territorial conflict lies in the colors blue and green, which, if exchanged between the two sides, could enable two states to live side by side in peace.

The Geneva Initiative has a map for each problem—and each solution lies in a powerful mix of colors. For the long-fought-over city of Jerusalem there seemingly is also a cartographic solution at hand (as revealed in Figure 6.8).

Plans for Jerusalem are predicated on the territorial divisions suggested in the Clinton Peace Plan put forth in December 2000,[10] which sought to have Arab areas within Greater Jerusalem, as well as within the Old City, become part of a future Palestinian state while Jewish areas were to be included within Israel. Moreover, Palestinian sovereignty was to extend over the Haram/Temple Mount, and Israel would be in charge of the Western Wall and adjacent spaces sacred to Judaism. Such divisions were to become subject to international monitoring.

[10] United States Institute of Peace (USIP). *Clinton Proposal on Israeli-Palestinian Peace, Meeting with President Clinton, White House*, December 23, 2000. Available at https://www.usip.org/sites/default/files/Peace%20Puzzle/10_Clinton%20Parameters.pdf. Retrieved January 25, 2020.

Again, the maps seemingly provided a visually persuasive and feasible solution to a long-standing problem and one that for most protagonists in the conflict seems insurmountable. The attractive colors of green and purple divided the urban landscape with an ease that made any such territorial agreement seem to reflect common sense. Indeed, as the cultural studies scholar Stuart Hall has already pointed out, politicians' most powerful tool is their appeal to what constitutes common sense—and common sense dressed up as cartographic artwork makes it all the more compelling and persuasive (McNally and Schwarzmantel 2009).

Zochrot

Among the array of Israeli NGOs there are also those that engage in "counter-mapping" of the state-defined territory. Counter-mapping is the act of making maps in cooperation with local stakeholders and community members in order to lay claim to land and propose alternative spatial visions, either past or present. Like their counterparts around the world, in Israel counter-mapping initiatives rely on new technologies ranging from GPS, OpenStreet Maps (an open source mapping platform), Google Maps, and satellite photography available through Google Earth. These technologies "remove 'oversight' from the sole control of authoritative professionals, planners, the military and the government and allow broad sectors of the population to read and understand maps and use them to create their own space" (Manoff 2011, 3). Counter-mappers can challenge and resist the state's hegemonic power to survey and map the land, and they can probe state-produced maps for the omissions and silences that were inevitably designed into them. In fact, indigenous communities in Australia have used such counter-maps as a kind of title deed and employed them successfully against the Australian state in a land ownership case (Crampton 2010a). Even if they don't lead to court cases, such counter-maps can provide an alternative spatial and political reality and attempt to remake geographical imaginations.

The NGO Zochrot was founded in 2002 with the aim of raising awareness among the Israeli public of the Nakba—the Palestinian displacement after 1948[11] and the large-scale destruction of the Arab topography that occurred in response to what in Israel is referred to as the war of independence. Palestinians started telling stories of the Nakba soon after 1948, yet it is only in the 1980s, with the rise of "New Israeli historians," also referred to as "post-Zionist historians," who critically examined the Zionist-informed historical

[11] For information, see Zochrot, http://zochrot.org/en/site/nakbaMap.

narrative of the establishment of the state and the role of the Israeli state in the displacement of Palestinians, that the issue started to emerge in public debates (Pappe 1995). Zochrot thus doesn't present a Palestinian narrative about the Nakba;[12] instead, it puts forward an Israeli-Jewish narrative in order to revise the dominant Jewish history of the state-making process (Lentin 2007). In other words, Zochrot's work is a form of "memory-work, which puts it on a collision course with the collective memory of Zionism" (J. Roberts 2013, 228).[13]

Zochrot organizes tours to Palestinian villages, sites, and towns destroyed during and after 1948. They collect, archive, and distribute material (including testimonies by people affected) about the Palestinian experience in historic Palestine that has been largely erased from the land, from Israeli culture, and from Israeli textbooks. Zochrot is thus encouraging an alternative historical memory and discourse. The organization's activities include producing alternative maps of the country that show the location of remnants of former Palestinian villages, towns, and sites that have been destroyed and thus erased from the Hebrew topography. Zochrot hereby challenges the dominant geopolitics of Israel's official maps and attempts to retrace the Arabic topography both before and after 1948. One part of the digital archive that Zochrot is putting together is a map of Israel onto which big round dots are superimposed, each representing a Palestinian locality that has been destroyed (as represented in Figure 6.9).

By clicking on the dot or by typing in the name of the Palestinian locality, users can access a Google satellite view of the area, its name, size, and history, including archival photos and testimonies. The map is a visually powerful digital archive of the Palestinian past.

Talia Fried (Fried 2012a, 2012b), who worked for Zochrot, argued that the criteria they used for including a Palestinian locality in a map was unclear as many localities had never been mapped before, and questions persisted about which localities to include (e.g., only those with permanent structures that were destroyed or also those that remained, but from which Palestinians had become refugees?). Moreover, depending on which sources are used—primary ones, such as oral and archival sources, or secondary ones, such as Benny Morris's *Birth of the Palestinian Refugee Problem* (1989),Walid Khalidi's *All That Remains* (2006), and Salman Abu-Sitta's *Atlas of Palestine* (2010)—the numbers of Palestinian

[12] For a Palestinian understanding of the Nakba, see the Palestinian London-based website https://www.palestineremembered.com/

[13] The power of the dominant Israeli-Jewish narrative is exemplified by the passage in the Israeli Knesset (parliament) of the "Nakba Law" in 2011, which prohibits funding to any institution or municipality that "commemorates the founding of the Israeli state as a day of mourning." The Nakba Law posed legal questions about civil liberties in Israel and the right of Palestinians to commemorate their history; but organizations such as Zochrot that were dedicated to commemorating the Nakba were potentially most directly affected.

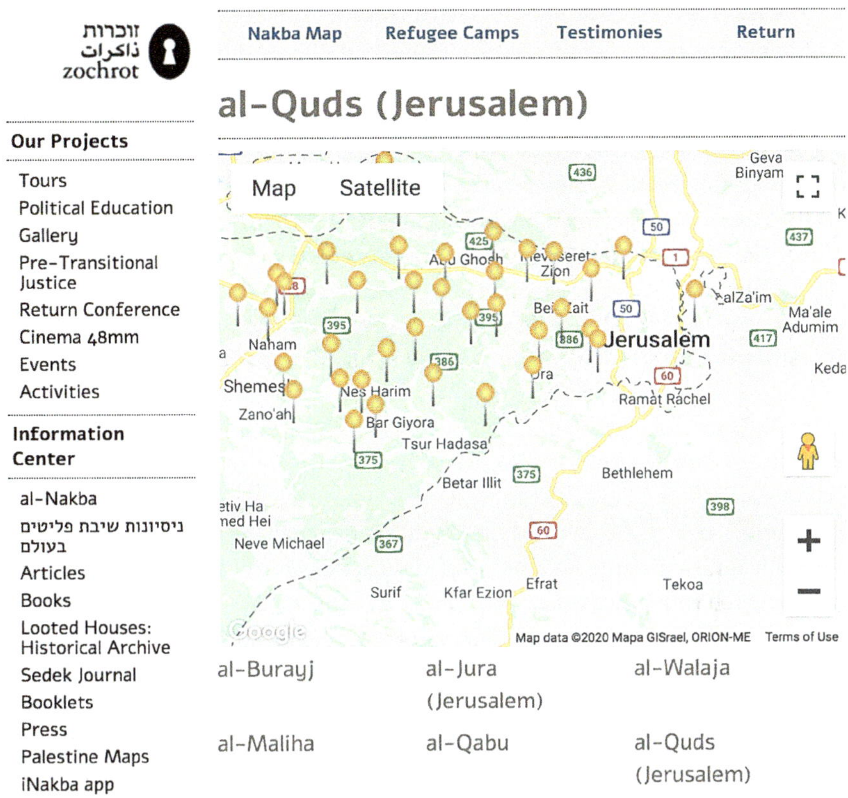

Figure 6.9. Zochrot website and mapping the Palestinian past in Jerusalem.
Source: Zochrot. Available at https://zochrot.org/en/village/49463. Retrieved January 25, 2020.

villages varies between 392 (in Morris) and 530 (in Abu-Sitta). The variation stems from different sources using different criteria for what counts as a village. According to Fried (2012a), "On Zochrot's map we listed all of the villages listed in Morris and Khalidi, and most of those listed in Abu-Sitta. We also added some that we came upon independently" (7) based on Zochrot's own redefinition of the Nakba as not a historically fixed but rather an ongoing event. What should be included or excluded from the map was determined by the sources used and by the map-makers' judgment.

Fried points out that such decisions "raise complicated questions about [the] moral imperative to remember and about the classification of trauma; on the other hand, Zochrot is aware that the map is a rhetorical and scientific instrument, it has the implied power to represent reality and representing this reality accurately is critical in a context where it is often denied" (Fried 2012a, 8). Therefore, mapping becomes a scientific, moral, and mythical enterprise.

Indeed, the appeal to science, objectivity, and historical accuracy is crucial for map-makers who attempt to gain legitimacy for any argument, whether textual or visual. At the same time, for Talia Fried, mapping is also a moral enterprise because as a result of Zochrot's mapping efforts, "The consensus of silence surrounding the Nakba has broken and on both the left and the right" (2012a, 9). But it is also a mythical enterprise. As Fried points out, "The classification that Zochrot came up with is also mythical. Or at least, it is highly symbolic and simplified. The dots are homogenous. They signify centers that have no real borders. All the dots are of the same weight. There is only one class of data by which to organize the geographical field (village name and location)" (2012b, 5). Maps therefore may help re-narrate the past and re-make "geographical imaginations" yet while doing so, science, myths, and morality always remain more intimately intertwined than the protagonists of pure science would have us believe.

The maps and map-makers discussed in this chapter represent diverse territorial visions. Some advocate drawing firm boundaries and establishing Israeli and Palestinian states side by side, and others challenge the dominant Hebraized geopolitical vision of the land of Israel since its inception in 1948 and demand that the Arab topography also get its due in history. The territorial consensus that has been fragmented since 1967 still gives organizations and activists who use new technologies and the internet to make maps the incentive to propagate their territorial visions to an ever-wider audience. Yet it is here where we may also encounter the limits of the democratization of mapping practices as maps produced by lay activists, who often lack socio-technical expertise and/or institutional legitimacy, may only reach no more than a committed in-group (Bittner et al. 2011; Bittner, Glasze, and Turk 2013). The persuasive power of the maps discussed here, however, stems from the very fact that they are more than just scientific maps as they constitute a pastiche of scientific and legal but also moral and mythical criteria; and at the same time, they are also representative of particular discourses that have become prevalent within Israeli society. The maps discussed in both this chapter as well as in Chapter 5, whether they were produced by right-wing or left-wing NGOs, point to not only the diversity of social movements that propel different geopolitical visions forward, but they also point to the increasing fragmentation of an Israeli populace that has become enveloped in a struggle that has gone far beyond maps and found its ways onto Israeli streets, where the political right is increasingly pitted against the political left (Chemi 2014). This increasing political fragmentation reveals the fissures and divides of the Israeli culture of consensus that once brought the Israeli nation into being.

7
Constructing Israel's Borders

Introduction: Why Boundaries Matter

Boundaries define nation-states. Ideally, they delineate sovereign control over territorial units, spatially delimit national control over land allocation practices, and separate nation-states from each other. At the same time, borders are not only material lines on maps or engraved in territories; they are also political, social, cultural, and discursive constructs. Border Studies scholars have increasingly focused attention on borders as sites where people produce and perform certain notions of territoriality, cultural assumptions, power relations, symbolism, border experiences, and social meanings (Brunet-Jailly 2011; Konrad 2011; Leuenberger 2011; Mezzadra and Neilson 2013; Paasi 2002; Paasi and Prokkola 2008; Newman and Paasi 1998). Such meanings are not fixed by the materiality of borders, but they are fluid, as they respond to the changing historical, social, and cultural contexts that shape cross-border relations and affirm or deemphasize the importance of borders.

Boundaries are particularly significant in the case of Israel, as many of its boundaries are contested and are still yet to be delineated. Post-1989 dreams of an open, borderless world are nothing but a fata morgana in the deserts of the Negev and the Sahara. In conflict-ridden areas, such as in Israel/Palestine, establishing firm boundaries so as to delineate territory and attempt to differentiate between ethnically and culturally diverse social groups remains a desired form of social organization (Brawer 2002). Historically, different Israeli as well as Palestinian visions of the territory—ranging from maximalist territorial dreams to more pragmatist accommodations to local realities—have meant that boundary-making is a deeply politicized issue. Israel's actual territorial limits and where the sovereignty of its state begins and ends is still uncertain (Miles 2011). Its borders continue to be a subject of dispute, negotiation, and conflict both within Israel and between Israel and its neighbors.

The emergence of borders in Israel/Palestine has a fraught history. During the British Mandate, the British and French delineated the territorial unit of Palestine based on imperial needs and power struggles. They mostly delineated its borders in Paris and London, with little or no knowledge of the local cultural, human, or natural topography (Biger 2002, 2008). Yet it is precisely that bounded entity of Palestine that remains contested to this day and that is often cartographically

The Politics of Maps. Christine Leuenberger and Izhak Schnell, Oxford University Press (2020).

DOI: 10.1093/oso/9780190076238.001.0001

claimed by Israelis and Palestinians alike. The way land is carved up on maps may be no more than an accident of history, but the consequences can reverberate for centuries, leading to strife, conflict, and bloodshed.

Despite Israel's still fuzzy borders, politicians and military experts have always been concerned with the need for defensible boundaries. They view the state's minimal hinterlands to be crucial for national security and for being able to defend the central area of Israel (often referred to as Israel's "soft under-belly"). Control over hills and crests, and the outward expansion of the boundaries have long been considered of strategic importance. Besides their presumed strategic military value, state boundaries also serve to integrate a polity by consolidating national identities. At the same time, they also became powerful dividers between supposedly different cultures. While Jewish culture and tradition has long served as a homogenizing agent, it is the bounded Israeli nation-state that also distinguished the Israeli populations from its neighbors. While the Israeli state, unlike, say, the French state, built a nation based on Jewish cultural communalities and a Zionist vision, it, much like the French "geobody" (Winichakul 1994) also came to include diverse social and ethnic groups ranging from Jewish Israelis to Israeli Palestinians, Druze, and Bedouins. At the same time, its boundaries powerfully shape cultural belonging and identity, and enhance "the otherness" of people across the border.

Historical events are a case in point. As part of the ceasefire agreement between Israel and Jordan in 1949, 30 Arab villages, located along the Green Line, were included in the Israeli territory in exchange for other territories. The Green Line served as a border between Jordanian-controlled territory (including what today constitutes the West Bank) and Israel from 1949 to 1967. Despite Israel's de facto annexation of the formerly Jordanian-controlled territory of the West Bank in 1967, making the border obsolete, cultural gaps continued to persist among members of the same clans on either sides of the boundary (Amara and Kabaha 1996). The boundary thus served to consolidate different social identities and increase cross-border differences. Given the importance of borders to state-making and nation-making, we now turn to how Israel's boundaries were made, mapped, and sustained, and how maps hindered as well as enhanced the boundary-making process.

Mapping Boundaries

Many actors have drawn and redrawn borders in the Middle East, ranging from the Ottoman Empire, the imperial powers of Britain and France, the Zionist movement, Trans-Jordan (later Jordan), Syria, Egypt, and the United Nations as well as Israel (Biger 2008). The lines they drew consist of four categories: "those in which natural features influenced their location, boundaries that run along old Ottoman Empire administrative boundaries, straight lines, new lines" (Biger 2002, 464). Natural boundaries, that follow geographical features such as a river, are considered the best possible boundaries. Ottoman administrative lines frequently have served as a starting point for negotiating final lines, as they set a precedent of where the lines might lie (such as between Israel and Egypt and Israel and Jordan). Straight lines (such as those separating Syria from Jordon and Iraq) are some of the most common boundaries in the Middle East (especially in desert regions), while new lines (including the line between Syria and Turkey) often follow a man-made feature such as a railway. Most of these lines were done with little or no consideration of local habitat or indigenous populations. Instead, many were delineated by imperial powers in Paris or London, that is to say, by experts who were far removed from the areas they so easily severed into two parts. It is only more recently that independent Middle Eastern states have redrawn their boundaries as part of bi-lateral agreements. Boundary-making in the Middle East is thus where top-down imperial aspirations continue to encounter attempts at local, bottom-up renegotiations of superimposed territorial lines. In the process, as maps are re-made and boundaries newly demarcated, people get reshuffled and identities, too, may get re-made.

Boundary-making entails a manifold set of technologies, techniques, and skills as well as a range of cultural and human resources. The long-term commander of the Israeli Defense Forces (IDF) mapping unit, Haim Srebro, argues that the process of boundary-making is comprised of boundary documentation, boundary delineation, boundary demarcation, and boundary maintenance (Srebro and Shoshani 2007). Various documents, including verbal descriptions, graphic charts or schemes, coordinates, and maps serve as resources to delineate boundaries. Yet the persuasive power of cartographically delineated boundaries fades in the light of demarcated boundaries. Existing markers and pillars along the boundary constitute more persuasive evidence for where the border might lie. Maps, then, are part of a more intricate system of objects, papers, and data that provide physical and conceptual resources to demarcate boundary lines "on the ground." At the same time, however, demarcated lines also need to be maintained and boundary markers regularly restored; granted, they may disappear, as they can be secretly moved, become eroded by harsh weather conditions, or buried in sandstorms. At the same time, because physical boundary markers need to be maintained, they retain their power only in conjunction with the written documents, maps, charts, and agreements that result from bi-lateral negotiations. Indeed, the jointly performed and precise documentation of the boundary in conjunction with the signing of an agreement on all the relevant documents pertaining to the boundary's delineation and demarcation, together make and maintain boundaries.

Maps, then, may not make boundaries, but they help co-produce them. While they are important ingredients for delineating borders, they become one of the many tools in the surveyors' and cartographers' toolbox. The map's level of accuracy and viability does not suffice, as its persuasive power derives from being part of an array of other documents and practices. This is partly because maps tend to represent information in ways that are most favorable to particular nation-states. Indeed, "the basic problem of maps is that they are usually published by one country and they are not common to both states. . . . [Therefore] states tend to produce maps that are in their favor" (Srebro and Shoshani 2007, 8). Moreover, maps also frequently employ different languages, names, and symbols; they rely on diverse coordinate systems, geodetic data, and cartographic projections and scales; so they frequently provide different interpretations of boundary lines.

Besides such non-standardized mapping practices across different cultural and historical contexts, cartographers also complain that maps provide weak evidence of boundary delineation as the quality of the mapping data, as well as the density, richness, and positional accuracy of geographical features, can vary across different maps. Maps therefore "are not reliable to be used as a sole evidence for the location of a boundary line" (Srebro and Shoshani 2007, 9); rather, they became part of a body of documents and data, demarcation pillars, and negotiation practices that can be used to ascertain geographical features and delineate a boundary line. How then have the many boundaries of Israel been delineated and which technologies and resources have contributed to their making? In the following we sketch out some of the conditions under which boundaries were delineated and demarcated and some of the real consequences such cartographic delineations have had on local topographies, people, and economies.

Delineating the Lebanese Boundary

The first international boundary between Lebanon and Palestine was delimited by the colonial powers of France and Britain in an agreement signed in Paris in 1923. The agreement comprised a description of the boundary and the location of boundary pillars and included three maps at the scale of 1:100,000 (1 cm on the map corresponding to 10 km on the ground) showing the pillars and features erected to delimit the area. The line produced, signed, and verified on a desk in Paris was to remain the line that would inform all subsequent negotiations of territorial sovereignty and belonging. Yet, in a war-torn region, lines, rather than remaining fixed, have a fluidity that alters in response to military campaigns and territorial advances by various protagonists at different time periods. It was only after the establishment of the Israeli state in 1948, its subsequent war with its neighbors, and the signing of the Armistice Agreement in 1949 between Lebanon and Israel, that the delimitation of their shared border areas commenced. A delegation of Lebanese and Israeli military generals, surveyors, and technical staff were to delineate and demarcate the border. The joint team surveyed the 1923 Mandatory International boundary as the basis for a new armistice line. After conducting joint boundary tours, making surveys and maps, and engaging in direct talks over the location of the mandatory boundary, the delegation agreed on many aspects of the line, yet some disagreements remained (Srebro, Adler, and Gavish 2009).

Where agreement could be reached, the boundaries were designated on Lebanese 1:50,000 (with 1 cm on the map representing 500 meters on the ground) scale maps and Israeli 1:20,000 (with 1 cm on the map representing 200 meters on the ground) scale maps. With maps in hand, Israeli and Lebanese surveyors, who were under the protection of a mixed Israeli and Lebanese military unit, were in charge of placing boundary markers along the newly delineated border. The plan was to designate additional boundary points in order to reach a total number of 115 signs (Szancer 2001). Demarking the line on the ground was vital given the discrepancies and lack of coordination between the Lebanese (originally French) and the Israeli (originally British) triangulation systems that were used to survey land, measure distances, and fix locations. The use of different triangulations networks could lead to divergences of several meters in the location of the boundary, which therefore made walking and demarcating the boundary all the more important (Gavish 2005).

Indeed in some areas, particularly along the Syrian frontier, the two sides did not agree on the location of the 1923 line, which resulted in diverging interpretations of the boundary's location. Such disagreements were aggravated by the fact that the French had not made cadastral maps of the land on the Lebanese side of the boundary, while the British had done so on the Israeli side. Worse still, the cadastral maps produced by the British had mapped land that the Lebanese claimed as theirs. Consequently, the two sides could not agree on the parceling of land and on how land ownership should reconfigure the new boundary line. With no agreement in sight, the team delineated three cartographic lines: the Israeli and Lebanese lines as well as the lines that followed Israeli land ownership claims. Irrespective of the three lines, subsequent allegations of land grabs continued on both sides. It was not until 1965 that a new delegation of Israeli and Lebanese surveyors and a UN observer attempted to re-delineate the border. The joint delineation, however, came to a standstill upon the commencement of the 1967 war between Israel and its neighboring states Egypt, Jordan, and Syria.

After the war, various attempts to map the territories continued until 2000. It was only that year that the then Israeli Prime Minister Barak decided to withdraw from Lebanese territories that Israel had occupied since the 1982 Lebanon War. In order to confirm Israeli withdrawal, the United Nations sent a team to the area so as to identify a line—known as the Blue Line—that was to conform to the Mandatory Line, yet it was not to be a formal border demarcation. In order to mark the line, the UN demarcation team had recruited surveyors that, according to a cartographer, "used old French maps and the old British maps and other evidence" (Interview B 2012) so as to newly demarcate the mandatory boundary. While both parties accepted it as a temporary separation line, it is not considered an international border agreed upon by both sides. Rather, it was determined by the UN's Cartographic Division in an effort to determine the mandatory international boundary with the latest technology available at the time (Srebro, Adler, and Gavish 2009). An Israeli surveyor and map-maker attests that such UN-defined lines are necessary when bilateral negotiations fail. It is in situations of conflict that the UN can become an "intermediary" and provide "a proposal or an intermediate solution"; yet such solutions remain "top-down" attempts, as they are not based on mutually agreed upon bilateral negotiations between the two parties (Interview S 2010). Thus, while the Blue Line prevails, it continues to do so at a cost. Indeed, the lack of a commonly agreed upon border gives rise to professional and civilian disputes. For instance, local Israeli military units often crossed the international boundary in attempting to control more territory to increase their tactical advantage. Such land grabs not only encountered vehement Lebanese opposition but also frequently pitted military units against professional Israeli cartographers who did not approve of these maneuvers (Interview Sh 2011). Moreover, the Blue Line also severed communities, like the village of Ghajar, which ended up one-third under Israeli control and two-thirds under Lebanese control.

For Asher Kaufman (2002, 2009) the partition of Ghajar resulted from various historical and cartographic errors. While the village had been under Syrian control until June 1967 and villagers identified themselves as Syrians, different maps from French, British, American, Russian, and Egyptian sources placed the village in either Syria or Lebanon or divided it between the two states. After the 1967 war, Israeli cartographers included the now Israeli-occupied Golan Heights, including Ghajar, in Israeli maps. They placed the Lebanon-Syria boundary line farther north, in accord with a 1963 Lebanese map. This exemplifies how maps and documentation, when used to construct a historical linage of a territory, can determine where boundaries might be drawn. Yet, during the entire process—whether past or present—the villagers were not consulted. Such top-down mapping practices continue to sit uneasily alongside the realities on the ground, and at times they can come at a high price. While maps distinguish between a northern and a southern part of Ghajar (naming its northern part al-Wazzani as in other previously produced maps), the residents of this one village (or two) regard their living space as one (Kaufman 2009). At the same time, although they hold Israeli identification, they remained disconnected from the state. More problematically still is the fact that as a result of the Blue Line, the village is considered a closed military zone with access restricted to its residents. Moreover, when Israel withdrew from its occupied territories in Lebanon in 2000, Hezbollah (an Islamic political party, the Party of God, that also is a paramilitary organization, formed in 1982, supporting Arab nationalism, Palestinian liberation, and frequently conducting guerrilla wars against Israel) took over the newly vacated territory, with the village becoming a hotspot for potentially violent flare-ups between Israel and Hezbollah. Consequently, with no bilateral negotiation over a boundary in sight and border tensions frequently erupting, the Israeli government was considering building a separation barrier along the Blue Line in Ghajar. The high price of failed negotiations and top-down mapping would once more be at the expense of the people of Ghajar. While the top-down mapping of the Israel-Lebanon border and the persistent lack of a binational negotiations continue to negatively impact people in that border region, the border between Egypt and Israel was one of the few borders its makers delineated with land and people in mind.

Delineating the Egyptian Boundary

By 1906, the Ottomans and the British delineated the border, also known as the Rafah-Akaba line. The 1906 line was the first fully demarcated border in the Middle East. It was also a "natural border" with its desert areas seen as among the best divides between two adjacent states. Also unlike, most border delineations, it was the first to take account of geographical features, water sources, and the population along the boundary line. In fact, at the time, the British carefully demarcated roads and water sources. Their attempt to survey and get to know the land, its water sources, and streets gave them a clear advantage when delineating the line (Brawer 1983). As the Ottoman Turks were not equally acquainted with the water sources in the area, upon delineation, all the water sources fell onto the side of British-controlled Egypt. However, as we have seen, demarcating a border is more than delineating a line on a map and surveying the land; it also includes erecting border posts. Indeed upon the boundary's demarcation on the ground, topographical and climatic conditions also impacted the final line, shifting it from the one delineated in maps. The rugged desert landscape made the poles that were to demarcate the line frequently invisible. Moreover, the British demarcation commission complained about the lack of water and the high desert temperatures and therefore wanted to finish the demarcation as quickly as possible. Thus, when the Turkish commission requested that the road from Egypt to Mecca be included on their side of the borderline, the British commission was more than willing to relinquish the road so as to finish the demarcation process without delay and escape the desert heat. Yet, despite the all too human factor that co-designed the 1906 line, the boundary continued to persist for years to come. After the collapse of the Ottoman Empire in 1918 and Britain's Mandate over Palestine, the same 1906 boundary delineated Mandate Palestine from Egypt, which was then under British sovereignty (Brawer 1983).

It was not until the establishment of the State of Israel and the subsequent 1948 war with its neighboring Arab states that the boundary with Egypt had to be newly delineated in 1949. Again it was the 1906 line that was to serve as the armistice line between Egypt and Israel from 1949 to 1967. In order to demarcate the boundary, a joint Israeli and Egyptian team of surveyors replaced lost border signs. In many border areas, they marked the boundary with barrels that were later to be replaced by concrete border posts. However, local farmers as well as military officers on both sides frequently moved the barrels so as to increase their land or improve their tactical position, respectively. Yet as the area was sparsely populated, only minor boundary violations occurred. The border with the Gaza Strip, though, had to be delineated separately following the 1948 war. According to the 1947 UN partition plan, that was to create independent Arab and Jewish States in former Mandate Palestine, Gaza was to be part of an Arab Palestinian state, yet Egypt occupied its territory after the war and considered Gaza under its sovereignty. The UN thus delineated this part of the border according to the position of the Egyptian army on November 13, 1948, and the Israeli-Egyptian Cease Fire committee approved the line on November 5, 1952 (Szancer 2001). During the delineation process, there was frequent turnover among the Egyptian experts, which slowed down the team's progress. Despite such setbacks, however, the Egyptian and Israeli teams' consensus on the boundaries' demarcation was ultimately made possible by the team members' commitment to professionalism, which enhanced the trust among them (Interview Sh 2011).

Yet the boundary as demarcated was not to last. After the 1967 war between Israel and its neighboring states Egypt, Jordan, and Syria, Israel more than tripled the land it claimed, as it occupied the territories formerly controlled by Jordan (West Bank and East Jerusalem), Egypt (Gaza, the Sinai Peninsula), and Syria (Golan Heights). Peace negotiations started during the 1970s, but it was not until after the 1979 peace treaty between Israel and Egypt that attempts were made to restore the 1906 mandatory line, with the exception of the boundary along the Gaza Strip. Surveyors encountered numerous obstacles in their task of re-delineating the boundary. First, they had to search for existing authorized and registered documentation, which was not an easy task as the 1906 agreement that specified the international boundary "was not well documented" (Interview S 2010). The team used those documents that were found and deemed reliable to help restore the boundary. The documents and maps, however, told many different stories about where the boundary might lie. As Haim Srebro and Maxim Shoshani (2007) point out:

> [According to an analysis of the] 1906 boundary between Israel and Egypt on 107 maps between the years of 1906 and 1982 [there] were . . . big differences due to a variety of reasons, like geodetic information, cartographic projection, accuracy of mapping features, accuracy of the boundary line interpretation and depiction, scale problems etc. The result . . . is that maps are not reliable to be used as a sole evidence for the location of a boundary line. (Srebro and Shoshani 2007, 9)

According to Israeli cartographers, then, maps constitute "complicated evidence," yet they also have "valuable potential in case that recognized coordinates and pillars do not exist" (Srebro and Shoshani 2007, 7).

In an interview, an Israeli surveyor who participated in the delineation of the border also points out that the "peace treaty was very lousy" and included "very poor, poor quality maps" (Interview S 2010). He complained that it included maps that were made not only without the involvement of Israeli experts, but that they also were not mapped onto either the Israeli or the Egyptian grid. Moreover,

> underneath the map itself you see just a white paper! Just a white paper! You don't have anything, no details. How can you come and use this in the field and then demarcate it by coordinates? But we don't use these coordinates, the Egyptians don't use these coordinates! So what can we do? Nothing. (Interview S 2010)

It is due to the maps' poor quality and small scale that the sand dunes had become invisible and represented nothing more than white paper. Consequently, at the suggestion of one of the Israeli experts, the team made ortho-rectified photo maps on a 1:100,000 scale, enlarged them, and transferred onto them the necessary lines and details. In the process, the team made sure that this was a "joint product." Indeed, as the interviewee attested, "Either we prepared them and they checked them—or they prepared them and we checked." Subsequently, both teams signed off on all the details, and only then did they have a "new original, which one can work with" (Interview S 2010).

Nevertheless, boundary disputes persisted over 16 border points, which were subsequently brought before an international tribunal. To solve the disputed points, the tribunal would often refer to boundary pillars that had been accepted by the two sides over time rather than relying on maps. It was not until 1988 that the international tribunal resolved the remaining territorial disputes. Not only did the available documents and maps complicate the newly initiated demarcation process, but the 1906 border, which at the time of its original delineation provided a model for border demarcation, now traversed a dramatically changed border area. Instead of only dividing large desert areas, it now also split highly populated Rafah. Yet, with maps, agreements, and jointly signed documents in hand, the line could finally be demarcated on the ground. The team went out in helicopters and "put the bars according to these auto rectified photographs" and thereby demarked more than 1,000 kilometers of land (Interview S 2010). No boundary, however, can persist without being maintained. This is especially the case where it cuts through a desert landscape, in which pillars can disappear among the shifting sand dunes. On the other hand, shifting sand dunes also could suddenly ease the surveyors' work. The surveyor recounted that while they were searching for "remnants of an old boundary pillar suddenly there was a storm—sand storm—and the dune, sand dune, moved and we found a remnant from 1906, under the sand dune!" (Interview S 2010).

While the Egyptian-Israeli boundary became an elaborate project that was co-produced by an aversion of the British to the desert heat, poor maps, contradictory documentation, and battles against shifting sand dunes, cartographers and surveyors agree that the newly delineated boundary, which relied on the latest technology and strictest guidelines as to how to come to a joint agreement, is now done "in a very exact way . . . by electronic means so you can clearly identify the exact position of the line—you can't move it even 5 meters in either direction" (Interview B 2010). Not only is it firmly embedded in border posts along the Egyptian-Israeli border, but its documentation through modern GPS systems affirmed its exact location and validity, and was signed into existence by experts from both sides on March 26, 1996. With that, the border stands, until sand dunes, a collection of maps, and some warring fractions again conspire to re-make and maybe even re-negotiate the line in the future.

Delineating the Jordanian Boundary

Unlike the 1906 border between Egypt and Palestine that was carefully surveyed and demarcated, the 1922 Mandatory Boundary that created two separate British mandates: Palestine and Trans-Jordan, was based on no more than a short verbal definition and was delineated only on small-scale 1:250,000 (with 1 cm on the map corresponding to 2.5 km on the ground) maps by the British governor Herbert Samuel (Srebro et al. n.d). The line was determined, as was customary at the time during the British Empire, according to natural geographical features. It ran along the center of the Arava, the Dead Sea, and the Jordan and Yarmuch Rivers. The area, however, was never surveyed, as the mainly desert landscape seemingly didn't justify such expenses. Consequently, the lines' geographical features were never adequately specified. The natural features along the line were neither fixed nor stable. The Arava experiences seasonal floods that could impact the prevailing topography; erosion of the Jordanian slopes could quickly complicate the determination of the boundary; and the Dead Sea and the rivers Jordan and Yarmuch also change their riverbeds and water levels over time, impacting measurements of a boundary line that was determined by its distance from the shore. As a result, the differences between the markings of the line on maps could reach up to 8 km (Srebro, Adler, and Gavish 2009).

It was not until 1946 that the Survey of Palestine under the British Mandate surveyed, marked, and documented part of the line, publishing two versions of a 1:250,000 map. After the 1948 war, Israel and Jordan signed an armistice agreement in 1949.[1] The agreement was to serve as a temporary ceasefire line to quell hostilities. It was not meant to prejudice territorial claims and rights in a final status agreement, nor was it to determine the status of the land that was to come under Palestinian control under the UN partition agreement of 1947. Indeed, the armistice line (the Green Line) was dictated purely by military considerations. While the northern and southern sections of the boundary followed the International Mandatory boundary, the central part of the boundary had been delineated along the ceasefire line by the Israeli military general Moshe Dayan and his Jordanian counterpart Glab Pasha. The hastily drawn armistice line (that divided what today is the West Bank from Israel) was to leave the boundary between the two regions ill-defined. To this day its location remains detrimental for some of the local population, as it cuts through villages, divides built-up areas, and severs villagers from their water sources and farmlands. Surveyors who subsequently tried to demarcate the line on the ground continued to re-assign units of land so the line would match up better with the cadastral map of land ownership. Yet the Green Line's delineation and status remains disputed even today. The story of the fraught Green Line that extended from Eilat in the south, to the centrally located Jerusalem, and Latrun, and on to the Gilboa Mountains in the north, took on a life of its own, apart from being part of the original delineation of Israel from Transjordan.

The 1967 war changed the territorial map yet again as Israel tripled the land under its control, including formerly Egyptian, Jordanian, and Syrian-controlled territories. It was not until 1994 that Israel and Jordan signed a peace agreement. Soon after that agreement was signed, a joint team of Jordanian and

[1] Ironically, the 1949 armistice agreements included an attachment of the 1946 edition of the map. Indeed, Israel continued to occupy parts of Transjordan in accord with the 1946 demarcation rather than the 1949 agreement. Thus, despite the 1949 agreements, almost all maps drawn post-1949 by both states show the boundary according to the 1946 interpretation of the mandatory international boundary.

Israeli experts began to delineate and demarcate the boundary line. As one cartographers recounts, during the peace negotiations,

> when the Jordanian side . . . saw that the Israeli side had brought a geographer and a geologist, they decided also to bring a geographer and geologist so then—at certain stages—the argument was not between the political negotiators but between the geographers on the Jordanian side and ours—and we agreed on matters—we said look—this is sensible—the Jordanians agreed to move the boundary in some places . . . to their side and we agreed, Israelis agreed, in other places to move. The actual boundary—that is now between Jordan and Israel—is not the boundary that existed at the time of the British. It is a new boundary, which was to a large extent the result of geographical considerations. . . . It is a much better boundary. A more sensible boundary. When the Jordanian side hesitated, or the Israeli side hesitated, it came to a direct contact between King Hussein and Prime Minister Rabin, and they were very lenient toward each other and very sensible. . . . [W]e worked it out and we showed each other the factors that should be taken into consideration, what would be reasonable and a favorable settlement on both sides, [which] put the boundary in a much better position. (Interview B 2012)

Another expert similarly emphasized that the development of cooperative relations between the two sides were crucial for successful negotiations over the location of the line. He maintained also that it was the professional surveyors' mutual trust and shared commitment to professionalism that enabled the Jordanian and the Israeli sides to solve practical problems and minimize disputes (Interview Sh, 2011). As a result, they agreed on a new line that cut through the Arava Valley, the Dead Sea, and the rivers Jordan and Yarmuk as well as through the Gulf of Eilat.

The teams also had the latest technology at their disposal, including orthophotos, satellite images, and satellite based surveys. They employed a joint geodetic reference system for determining coordinates of the line. For instance, in order to delineate the marine boundaries along the Dead Sea and the Gulf of Akaba, the Global Positioning System (GPS) was used to locate six reference points on the Israeli side and six on the Jordanian side that enabled the boundary to be located independently of topographic characteristics, which, as noted, can shift and change over time, yet relevant geographical features were taken into account along the proposed line. Thus while the 1922 Mandatory Line as well as the 1946 demarcations done under the British Mandate served as points of orientation, the new line does not concur with them any longer. While the surveying team used various documents, including maps, according to one surveyor,

> There was a problem which maps would we use. Then I prepared two versions. One thing was . . . which language [to use]? Which names? Because they sometimes have for the same areas other names. So what I suggested was—on the Israeli side—we use our names. On the Palestinian side we use their names. (Interview S 2010)

Despite the production of new maps, the team decided that the delineation's geographical coordinates, along with the boundary pillars (which were to be regularly maintained), were to take precedence over maps during any future negotiations (Interview S 2010). Placing boundary pillars was therefore a high priority. Yet the difficult topography and the possibility of mines that had been swept into the valley over the years made positioning them no easy task. Nevertheless, the Israeli-Jordanian team placed 124 boundary pillars from the Gulf of Eilat up to the saltpans of the Dead Sea. Subsequently, the surveys and documentation pertaining to the delineation and demarcation were signed and adopted by both sides and were post factum adopted as part of the peace agreement. As the participants proudly point out, the Jordanian-Israeli peace agreement included a new model for delineating boundaries that exemplified good practice, as it documents the process of the agreement, includes all relevant documents and surveys, and specifies how to maintain the boundary. Indeed, the joint team of experts works to this day to assure the delimitation and maintenance of the boundary. As one team member pointed out:

> For the last 16 years, until 2, 3 months ago, we shared this post. I'm still the chair of the joint team of experts with Jordan. He [his Jordanian counterpart] left three months ago, because he was to be the director of surveys over there; too busy! Somebody else replaced him. This is very important, because since the peace treaty in 1994, we prepared the delimitation of the boundary. (Interview S 2010)

The Jordan–Israel boundary serves as a model for two reasons: First, according to protagonists, experts and surveyors used the most up-to-date technologies, such as orthophotos, satellite images, and satellite based surveys, to specify the boundary so precisely that it is considered to be the most exactly demarcated boundary in the world today (Interview B 2010). Second, it exemplified how boundary-making is an interactional achievement. For the boundary between Israel and Jordan to come into being, Jordanian and Israeli teams closely cooperated over extended periods of time. Boundary engineers, survey personnel, and the highest political authorities, including King Hussein and Prime Minister Rabin, were involved in the process. Indeed, this was the first time that Israel had conducted official negotiations with an Arab country over delineating a boundary (Israel had conducted talks with Egypt over demarcating the boundary but not over delineating it.) Accordingly, it was the joint precise documentation of the boundary that crucially contributed to its stability (Srebro and Shoshani 2007). It is thus not just technologies but also the joint cooperation, face-to-face contact, and painstaking work of combined Jordanian-Israeli teams that produced a boundary that to this day remains stable and free of controversy.

Yet, despite all its advantages, the question of boundaries between Israel and the former area of Transjordan remained controversial, due to the Israeli government's legal interpretation of the status of the Palestinian Territories. For protagonists from the government's official mapping institutions, the Green Line has been replaced by the 1994 peace agreement between Israel and Jordan; therefore, they maintain, the 1949 armistice line is obsolete. Under international law, however, the Palestinian Territories remain under Israeli occupation (after having been illegally annexed by Jordan from 1950 to 1967), as the West Bank had been assigned to a future Arab state under the 1947 UN partition plan. Hence, such different legal interpretations of the territorial status of the area have contributed to continuing territorial disputes. The Jordanian-Israeli border has not solved the rights of Palestinians to their own territory according to international law. The story of the 1949 armistice line, therefore, not only pertains to relations between Israel and Jordan but also to Israel's relations with the Palestinians: a story that yet has to be told.

Delineating the Green Line between the West Bank and Israel

The story of the infamous Green Line is the story of a blue pencil. Only after its delineation was the line overprinted on Israel's official maps with the color green, providing it with the name it has today (Brawer 2002). After the 1948 war between Israel and its Arab neighbors, the line between Israel and Transjordan was to be delineated during the armistice talks between Israel and Jordan in 1949 in Rhodes. As we have discussed in the previous section, this meeting was attended by the Israeli military general and head of the Israeli mission Moshe Dayan. As one participant in the meeting recounted:

> Either purposely or because he [Dayan] was not a very good map-reader, he took a blue pencil and he drew the line and the Arabs were less [qualified] cartographers than he was. Instead of cartographers to draw the line, they drew the line. And he took a blue pencil and he drew the line and the width of the line of the pencil was nearly 2 millimeters—being [equivalent to] nearly 300 meters—so you see if you take one side of the blue line or the other side you have two different boundaries. And I was present there . . . and I was at the time

> still a graduate student . . . and I got their attention. . . . I said, you know what you are doing? . . . [T]he line itself is 300 meters! But he dismissed it. (Interview B 2012)

This unreflective dismissal was to come with huge costs. Not only does the thick blue line mean that "there is a swath of country which is 200 or 300 meters wide that is disputed as to the exact position of the line" (Interview B 2012), but as a cartographic expert points out, there are

> actually three lines—we have the Israeli Green Line which includes Eastern Jerusalem and some minor other areas, we have the Jordanian or the Arab Green Line, which actually concurs with the armistice agreement between Israel and Jordan, and we . . . have the Green Line of the United Nations, which are the maps of the United Nations truce supervision body, and they are not identical! (Interview B 2010)

The UN maps were based on the armistice agreement and amended according to the UN Mixed Armistice Commission (that operated between 1949 and 1967), which included Israeli and Jordanian military representatives and which was chaired by a UN officer. Whenever there were incidents along the boundaries, the commission addressed respective Israeli or Jordanian complaints and voted in favor of one or the other side. Although the commission's decisions were documented, they were not implemented and amendments to the line were not executed (Interview B 2010). Not only was the UN interpretation of the line not implemented, but the Jordanians also never demarcated their line. The Israelis, however, soon after the signing of the armistice agreement, went out and implanted their line on the ground with boundary posts, barrels, and heaps of rocks.

> Israel demarcated the line and they invited the Jordanians to participate in the demarcation. They refused. What they did, they sent observers, which were not members of the teams which demarcated the line, but they were . . . present nearby and they had a look from the other side. . . . In some cases, they protested "this is not the correct line," but their protests were ignored. Israel demarcated the line. . . . [T]hey put in various forms of boundary posts. . . . The Jordanians had their own maps. . . . [O]n the maps they had their version of the position of the line, but they did not demarcate it on the area itself as the Israelis did. (Interview B 2010)

Thus with three Green Lines, yet only one of them demarcated, agreement on the line's location was hardly likely. As one expert points out:

> To put it bluntly the Israelis are very determined that their line is the correct line and they will not negotiate on any other line. The Palestinians who inherited all these documents from the Jordanians when Jordan withdrew from Palestine, are not very sure about their line. . . . They have their doubts, and in some cases, they don't have one version. (Interview B 2010)

With the Palestinians unsure of their line and Israelis convinced of theirs, the United Nations brought their maps to their New York headquarters and restricted their access. Indeed, these maps are now "absolutely confidential. . . . The United Nations felt that these documents may either help the Palestinians or the Israelis and they don't want to take sides. They want to be completely neutral" (Interview B 2010). As the armistice line was never fully demarcated and does not follow any physical features, there is now a frontier zone of about 5 km that is ill defined, yet severely impacts the local population (Brawer 2002).

With the line in place, however fuzzy its boundaries, it severely impacted cross-border regions, economic development, agricultural practices, and social interconnectivity across the border region. After 1949, Arab villagers who by and large worked the agricultural land in the lowlands lost access to their farmlands, employment opportunities, markets, and services. At the same time, the Israeli government sponsored the establishment of Jewish settlements along the border. Settlers, who often consisted of new Jewish immigrants, were—like early Zionists—to redeem the land, implant Jewish culture into its soil, replace depopulated and displaced Palestinian communities along the Green Line, secure Israel's new frontier region, and in doing so participate in the mission of nation-state building and its expansion into the newly acquired territories (Falah 2002; Hasson 2012). Such spatial developments and government-sponsored resettlement programs led to far-reaching changes in the human and natural landscape of the frontier. While the Jewish population increased on the Israeli side of the line, the Arab population decreased on the Jordanian side. At the same time, ethnic and cultural identities on both sides become more and more distinct. Arab villages on both sides of the borders which, before 1949, had similar geographical and socio-cultural characteristics become increasingly dissimilar once divided by a border (Brawer 1978). Economic activities and modernized agricultural techniques made the Israeli frontier ever more distinct from the Jordanian border region. The very different topographical development on the two sides become so stark that on air photographs, it was possible to identify and locate the boundary line even in areas where the boundary was not demarcated (Brawer 1978). The new border thus led to a bifurcation in economic and topographical development as well as separating two increasingly contrasting political and administrative systems. With the border providing an ever sharper division in the natural and human landscape, tensions also rose along the frontier zone.

As noted in an earlier chapter, after the 1967 war and Israel's de facto annexation of the West Bank, the Israeli government decreed that the Green Line was no longer to be represented on official maps printed for the public. Yet the line continued to appear on administrative maps. In other words, "in the West Bank there are no Israeli recognized sub-districts, so the boundary of the administrative division is actually the Green Line" (Interview B 2010). Moreover, maps used by the military continued to display the Green Line. Crossing the Green Line onto the Palestinian side, no matter whether it exists cartographically or not, entails not only entering a territory subject to different legislation, governmental services, and developmental policies, but also an area that is governed by Israeli military law. Therefore "virtually—the Green Line exists. It exists in practice, in a good many ways" (Interview B 2010). Despite its administrative significance, not to print the Green Line on official maps, however, has its politics:

> So practical maps on which the administration operated had the Green Line. But for public use, maps for schools, wall maps, public places, no Green Line! The basic view was that this is not a boundary. This is only an agreed temporary separation line and therefore we don't want to publish it and by thus recognizing that this is the boundary of Israel. (Interview B 2010)

With the official elimination of the Green Line, other changes transpired as well. Indeed, the Green Line's border function of separation was virtually eliminated overnight. Economic exchange increased yet again, with ever more Palestinians seeking employment in Israel proper. At the same time, with the Green Line no longer being the frontier zone, Jewish settlements moved into the heart of the West Bank. As a result, Jewish and Palestinian enclaves developed side by side yet separate from each other. Their proximity only accentuated their differences. The ethnic, religious, and social characteristics as well as the standard of living within the Jewish enclaves existed largely in discord with their cultural environment. At the same time, they remained completely dependent on Israel (Brawer 2002).

The expansion of economic interchange quickly came to a halt with the start of the first Intifada in 1987. As a result, the Israeli government reimposed the Green Line's function as a full-fledged border. Economic exchange was minimized and the border became increasingly closed for Palestinians attempting to enter Israel. Foreign workers from Thailand, China, and Romania quickly replaced the former Palestinian workforce inside Israel. Until 1993, economic links between the West Bank and Israel were at a virtual standstill, with severe economic consequences for West Bank Palestinians. It was not until the 1993 Oslo Accords and the end of the Intifada that economic relations improved.

Besides the subsequent increase in economic interchange, cross-border criminal activities, such as the stealing of Israeli cars for delivery to Palestinian-run "car slaughterhouses" (where cars were disassembled and their spare parts resold) was also thriving. Yet the thaw of cross-border relations, whether legal or illegal, was again short-lived. At the end of 2002 and the rise of the second Intifada, the legal movement of a Palestinian labor force into Israel declined once more to a trickle, with again devastating effects on the Palestinian economy. To this day, the Israeli government may have erased the Green Line from official maps, but it has continued to enforce the line's existence for different social groups in diverse ways. While Palestinians are acutely aware of its existence as for them it presents a strictly enforced border, Israeli settlers and military personal can easily and often unknowingly cross the line. The Green Line, then, is founded on a radical "provisionality" that continues to leave questions of borders, sovereignty, and polity open-ended and unsettled.

The Green Line today is even fuzzier than when it was first drawn as a blue line in 1949. According to an Israeli cartographer, a mutually agreed upon borderline is as far from Israel's consciousness as ever. Indeed he maintained (in 2012) that his government does not "want to hear anything about it. . . . [T]hey don't want to enter into any discussion of the question of the boundary" (Interview B 2012). At the same time, Palestinians have refrained from proposing possible territorial divisions as "any new proposal would jeopardize what they have achieved in the UN and could give Israel the wrong signal—namely that Palestinians might agree to live with less territory" (Falah 2002, 506). However, one of Israel's primary cartographic experts, who was also consulted by Arafat's advisors, would tell everyone, including his high-ranking Palestinian visitors as they sat in his office, "to me the Israeli [version] isn't accurate, the Arab version is not accurate and the UN version is not accurate, and not only this—they are bad!" (Interview B 2012). The line also has many faults that are "detrimental to one side or the other" (Interview B 2012). There is no doubt in his mind that for any future peace agreement, the line has to be drawn anew. While changes would not be extensive, they nevertheless would have to take into account geographical matters, assure that villages were not split into two, and be drawn so that communities were not cut off from their water sources and farmlands. Exchanges of land to improve the line for both sides, however, would depend on "goodwill and good sense," and on a mutually satisfactory "agreement on how to administer the boundary and the agreement of good neighborly relations" (Interview B 2012). Therefore, a boundary cannot function simply as a result of being demarcated and maintained; its sustainability crucially depends on mutually satisfactory agreements of what constitute good neighborly relations that define the parameters for cross-border economic exchanges and social mobility.

Delineating the Syrian Boundary

While the Green Line is "bad" (Interview B 2012), the Syrian boundary doesn't fare much better. According to the Sykes-Picot Agreement in 1916, the United Kingdom and France divided the Middle East in terms of their respective spheres of influence and control. Their plan was to be implemented on the defeat of the Ottoman Empire during World War I. While the region of current-day Israel, Jordan, and southern Iraq was to be under British control, part of Turkey, Iraq, Syria, and Lebanon were to come under the French sphere of influence. The imperial powers were also to delineate state boundaries in these regions. Yet the delineation of the boundary between British Mandate Palestine and Syria took Britain and France nearly ten years, in doing so they took account of all relevant agreements, including the Sykes-Picot agreement. They also relied on various historical maps in order to facilitate the demarcation process. The subsequent line was delineated in Paris and London and its architects had little or no knowledge of the area. The making of this map was thus far removed from local realities and the needs and practices of the affected population. It was not until 1921, that a survey team found that the delineations did not correspond to village boundaries. Consequently, after surveyors consulted with heads of local villages, they re-drew the lines in accordance with the newly acquired information. Their work, however, was not trouble free. Indeed "the work of the demarcation commission was interrupted by raids against French Syria conducted by Syrian exiles" (Biger 2002, 468). Yet, a new line ultimately emerged; units of land were transferred between the two sides; and by 1923 the international boundary between Palestine and Syria-Lebanon came into being. Subsequently the 1926 Bon Voisinage agreement between the High Commissioner of Palestine and of Syria spelled out regulations concerning land cultivation rights, movement and trade across the newly established border (Biger 2004).

But this was not to last. After the 1948 war, an armistice agreement between Syria and Israel was signed in 1949, with the ceasefire line delineated on a 1:50,000 map. Even with the map being made to a scale that was relatively fine, Syrian and Israeli experts had to deal with a range of boundary disputes and hostilities over land claims and water sources. Relations were strained between the two sides, which impacted their ability to meet face to face and address territorial disputes, and it hindered the production of a joint boundary delineation on 1:20,000 maps (with 1 cm representing 200 meters on the ground) which could have addressed local needs and land claims. The final agreement specified that the line parallel the 1923 international boundary, which had been originally delineated by the imperial powers. Yet as the international boundary and the ceasefire lines did not concur, coordinates of the demilitarized zone kept shifting.

Given the lack of coordinated progress on boundary delineation, Israel decided to delineate the boundary unilaterally. The one-sided demarcation project did not go unnoticed, however. The missions of Israeli surveyors within the Demilitarized Zone (DMZ) and their presence along the disputed boundary, often in the dark of night, were carried out under threat of enemy fire (Interview Sh 2011). Their penetration into the DMZ were done under the protection of the Israeli military in order to secure the surveyors' mission. Despite these hazards, the Israeli government was committed to delineating the border, as this was considered crucial for fostering development along the boundary area. By 1951, Israeli surveyors demarcated the boundary along the Dan River (the largest source of water entering the Jordan River). By 1958, they demarcated the boundary along the Hula valley in order to then drain the swamps and develop a national water system (that transported water from the northern part of the Jordan River through the lake of Galilee to the Israeli south). The Syrian government opposed Israel's development projects on the grounds that they interfered with the DMZ's status quo and violated the ceasefire agreement. While the UN gave some credence to the Syrian position, they called on both sides to resolve the dispute (Shalev 1989). Israel, however, maintained that the agreement only prohibited military activities and that development projects did not change the military balance in the region. They also justified the Hula project by arguing that Britain had planned to develop it as early as the 1930s. Yet, despite such attestations to the project's non-military nature, when conflict between the two sides erupted again, Israel used the developed infrastructures for military purposes. Indeed, such projects enabled the military establishment and the Ministry of Foreign Affairs to affirm Israel's sovereignty over disputed territories, which, along with their economic benefit, were considered vital to national interests (Duani 2010).

The boundary, however, was to be redrawn yet again. After the 1967 war, Israel occupied the Sinai and the Golan Heights amongst other territories, and by 1973 a coalition of Arab states conducted a surprise attack on Israel on Yom Kippur (known as the Yom Kippur War), with Egyptian and Syrian forces crossing the ceasefire lines. After the 1973 Yom Kippur War, Israel again made territorial gains, signing a disengagement agreement with Syria. The Survey of Israel, with assistance of the UN, was to demarcate what was to be the Separation of Forces Agreement in 1974. But it was not until 1995 that the two sides discussed future boundaries as part of peace negotiations. While the negotiations continued until the year 2000, they led nowhere. Despite accumulated cartographic expertise and advances in techniques for boundary delineation, the lack of agreement and cooperation between the Syrian and Israeli sides have continued to hamper any progress toward delineating a mutually agreed upon boundary. Indeed to this day, Israel is on both sides of the armistice line as

"Israel occupies the Golan Heights—[which] according to the armistice line is part of Syria" (Interview B 2012).

Boundaries on Maps

As we have seen, Israel's boundaries have long been borders in flux. How then has Israel's official mapping institution—the Survey of Israel—represented these lines on its maps? In the remainder of the chapter we highlight some of the Survey of Israel's commonly used mapping strategies (Miles 2011). Israel's boundaries are so numerous and manifold that the *New Atlas of Israel* (Survey of Israel 2009) reproduced a map specifying six different boundaries statuses, including mutually agreed upon International boundaries (such as the Israeli-Egypt and Israeli-Jordan borders); boundaries (such as between Israel and Jordan) that do not include specifications as to the status of the territory on one side of the boundary (as is the case in the West Bank); borders (such as between Israel and Lebanon) that have been agreed upon by Israel and the United Nations but without the approval of the Lebanese government; ceasefire lines (such as between Syria and Israel) that can, upon lack of agreement, lead to a demilitarized zone; and maritime boundaries (such as the boundary along the Red Sea (as specified in the Israel-Jordan Peace Agreement). And the 1993 Oslo Agreement with the Palestinian Authorities, with its division of the West Bank into Zones A, B, and C, exemplifies yet a different form of border-making. Thus, while Israel's boundaries have stabilized over time, the temporary status of many is also highlighted. Indeed, according to the *New Atlas of Israel* (Survey of Israel 2009) ceasefire and armistice lines are based on armistice agreements, which enables each side to the conflict to ultimately claim different rights and positions within the framework of peace negotiations (Srebro, Adler, and Gavish 2009). With only two boundaries based on peace treaties (with Jordan and Egypt), most, therefore, are still to be defined and remain a matter of future peace negotiations.

Yet with the potpourri of maps that could be presented in an atlas, some maps are systematically highlighted whereas others end up omitted. It is notable that throughout the history of Israeli map-making, official maps included the imperial boundaries delineated by France and Britain unless such lines were replaced by new international boundaries that resulted from peace agreements (with Egypt and Jordan) in conjunction with the continuously to be revised ceasefire lines. Even when they did not coincide with the ceasefire lines, imperial boundaries were there to stay; the 1947 UN partition plan was short-lived in comparison. While both sets of boundaries, as depicted on imperial maps and the UN Partition Plan, can be understood to legitimize the establishment of the Israeli state (as British Mandate maps incorporated the 1917 Balfour Declaration supporting a national home for the Jewish people in Palestine and the 1947 UN Partition Plan proposed the establishment of both a Jewish and Palestinian state in Palestine), the absence of the UN partition maps in favor of the imperial boundaries is notable. Indeed, "states tend to produce maps which are in their favor" (Srebro and Shoshani 2007, 8). As a case in point, Israel's official mapmakers delineate boundaries that favor Israeli territorial claims and therefore tend to exclude the boundaries proposed in the UN Partition Plan which point to Israel's annexation of territories assigned to a future Palestinian state.

On the one hand, protagonists from the Survey of Israel argue that they only include those boundaries that possess official status and are practically relevant at any given time. In doing so they present their actions as beyond human judgment and informed solely by the empirical characteristics of the natural world. On the other hand, they often readily admit to the "contingent" nature of their practices. As they point out: cartography hinges on "the decision of a cartographer" (Interview S 2010).[2] As one eminent cartographer noted:

> The scholarly approach to cartography is to give a true and unbiased picture of what exists in the area. Now—you must be subjective in that respect. . . . [E]veryone has a different view on what unbiased means. So . . . cartography, even if fully subject to geographical considerations is to a certain extent subjective, because the editor has his views and visions on what matters and what is important. . . . So all cartography has a certain minimal subjectivity. You cannot escape that. (Interview B 2012)

It is not only which features, names, languages, and sections of a country are included in a map that can become politics by other means, but a sequence of maps that are there to tell the history of a state can also tell that tale by virtue of the omission or inclusion of select maps.

Official Israeli-produced maps of the country tend to emphasize not only the temporary status of boundaries, but they also favor the representation of Israeli territorial claims. At the same time, while the Israeli conquests of what are considered Palestinian territories is deemphasized through elimination of the UN Partition Plan map, maps do tend to highlight Egyptian, Jordanian, and Syrian territorial conquests as infringements on Israeli land claims, even though these

[2] Cartographers and surveyors from the Survey of Israel are thus not unlike other scientists when referring to the nature of their scientific practices. Gilbert and Mulkay (1984) found that scientists use an "empiricist" as well as a "contingent" repertoire when referring to the validity of their science. According to the "empiricist" repertoire, scientists' actions are presented as following unproblematically from the empirical characteristics of the natural world. Under the "contingent repertoire," their personal judgment and circumstances are presented as impacting their science.

territories were to be assigned to a future Palestinian state. As was pointed out in the *New Atlas of Israel,*

> During the war [1948], areas of the land of Israel were conquered: the areas of Judea and Samaria by Jordanian and Iraqi forces, the Gaza strip by the Egyptian forces and along the Jordan River and the Sea of Galilee by the Syrian forces. (Srebro, Adler, and Gavish 2009)

To be sure, the very terms used to describe the territory, such as Judea and Samaria, rather than West Bank, indicate that the land is claimed as part of Greater Israel. This nationalist expansionist ethos becomes one of the underlying assumptions that inform the language and terms used to talk about land, territories, and boundaries. The state's official map-makers thus came to represent their state's vision. As one cartographer noted:

> I know I am responsible for producing reliable products. . . . I am the director of [name of institution]. [Name of institution] is the scale of the state. The ruler of the state. And we have to be correct and precise. (Interview S 2010)

Accordingly, what is considered professional is also considered good science, and it is good science in the service of the state.

While a state's institutions, such as its official mapping agency, have to make politically sensitive decisions concerning boundary delineations and how to represent the status of certain territories, those we interviewed who had been part of this process insist that they generally experienced no direct political intervention. Indeed, direct political intervention is rare. The decision by Yigal Alon, minister of education after 1967, to eliminate the ceasefire line with the West Bank from official maps for the public was not the norm. Indeed, mostly it is the staff cartographers themselves who weigh professional considerations, their responsibility as representatives of the state's vision, and their interest in producing maps that are uncluttered, visually appealing, and representative of relevant features on the ground. While they are recruited to further a state's political goals and interests, they also use their autonomy to shape the very implementation of such goals and interests (Kimmerling 1998). Just as in other institutional contexts, actors do not have to follow explicit guidelines or political directives; rather, hegemonic cultures within institutions are informed more by what is considered both "good practice" and what constitutes appropriate cartographic style and language. It is thus informal agreements, tacit knowledge, and impromptu improvisations that are guided by a commonly shared vision. That vision is further strengthened by the close historical interlinkages between the Israel's state cartographic division and the military mapping unit.

As we saw in chapter 3, the interlinkages between civilian and military mapping have long served cartographic developments in Israel—not least its ability to wage war in 1948 with the best cartographic knowledge and expertise available at that time. Civilian and military mapping efforts were thus combined to achieve commonly shared nationalist goals and strategic advantages.[3] Even under the British Mandate, civilian mapping units worked side by side with British cartographic experts so as to enhance military objectives. After the establishment of the Israeli state, surveyors and cartographers took part in peace negotiations as well as the delineation of ceasefire lines. As a matter of fact, during the 1950s and 1960s the Survey of Israel's cartographers participated in military activities, while members of the military subunit participated in delineating boundaries for peace agreements. Moreover, during the boundary demarcation with Syria, civilian surveyors were willing to enter the DMZ to survey and demarcate Israel's border even under threat of enemy fire. In this case, whether engaged in civilian or military cartographic and surveying activities, the Israelis involved could be from either the civilian or military mapping unit. Such formal interlinkages and their resultant informal networks between the civilian units to the political establishment blurred the borders between cartography for civilian and military purposes and sustained the hegemonic politics of a system, the seeds of which went back to the early establishment of the Yishuv (Kimmerling 2001).

As we have seen in this chapter, the history of Israel is also a turbulent history of boundary making and its failure. At a time, when transnational entities such as the European Union (EU) and North American Free Trade Agreement (NAFTA), as well as global trade and cross-border movement, have seemingly replaced the 19th-century enthusiasm for Westphalian concepts of nation-states, in the Near East, attempts at state-making and boundary-making in the shadow of imperial politics and arbitrary boundary delineations are still reverberating into the 21st century. In this region, boundaries have made nations yet also precluded cohesive social groups from striving for nationhood. Indeed, artificially delineated imperial boundaries, whether they differentiated Egypt from Palestine, Lebanon from Jordan, or Syria from Iraq, become means to create nations and a people who identify themselves with that nation. Yet many social groups, such as the ethno-religious Druze, that have a long-standing, commonly-shared cultural identity, were never offered the possibility of forming a nation (Biger 2002).

[3] This is not to say that there were no exceptions. As our interviewee points out, the withdrawal from Lebanon in 2000 and the delineation of the Lebanese-Israel boundary was a purely military affair from which the civilian unit felt excluded after the civilian subunit within the Survey of Israel become institutionally separated from its military counterpart (Interview Sh 2011). Indeed, he maintained that in this case the military's wish to control the delineation and withdrawal and further nationalist goals clashed with the civilian unit's dedication to professional integrity and careful cartographic considerations of the boundary line.

The Zionist nation-building project, too, was fostered by the 19th-century assumption that populations desired to be part of a nation-state that consisted of an ethnically and culturally homogenous group that differs from its adjacent nation-states. Indeed, the Israeli state has historically implemented policies that would assure an ever more ethnically and culturally homogenous Jewish state. Just as the establishment of the French state paved the way to the formation of the French people (Hobsbawn 1992), the establishment of the Israeli state and its borders has also assured that its population would become ever more homogenous and culturally uniform.

Besides the role of borders to help create a state and a nation, their very delineation reveals the all too human stories that inform their making. While experts point to the latest mapping technologies, standardized coordinates, and boundary-making techniques as ways to exactly and accurately delineate boundaries, it was always and will always also be interactional contingencies, preferences, likes or dislikes, and trust or distrust between negotiating parties that will crucially impact whether boundary delineations become a success or a failure. Just as the aversion of the British to the desert sun in the Sinai helped co-construct the international border between Israel and Egypt, the inability of Syrian and Israeli experts to find common ground over how to delineate a commonly shared boundary line became a deeply interactional affair. There is no technical solution to boundary-making and peace, however appealing such an approach may seem. Good boundaries are based on attending to the needs of local populations, negotiating with the experts, policymakers, and politicians on the other side, and establishing relationships based on trust, professionalism, or commonly shared guidelines. Ultimately, it is these relationships that enhance what one cartographer called "goodwill and good sense" (Interview B 2012) between the two sides, which makes peace ultimately possible.

8
Map-Making for Building the Palestinian Nation-State

Introduction: Enemies Closer

The Israeli nation-state building process must be understood in the context of the Israeli-Palestinian struggle over land, legitimacy, and history.

> Israeli state- and society-building were not solely an internal Jewish affair. In fact, the distinct characteristics of the Jewish-Palestinian conflict influences and decisively shaped the character of the Jewish state-to-be and continue to do so. (Shafir 1999, 82)

Accordingly, the Palestinian-Israeli conflict is not external to the Jewish project of establishing a state, but rather, according to Uri Ram, "the axis of the conflict between it and the indigenous population shaped the structure of the emerging Israeli society" (Ram 1999, 64). Thus, it is profoundly implicated in the formation of a hegemonic and collectivist Zionist culture in both pre- and post-1948 that defined itself against the "indigenous other." Consequently, the geopolitics of the region was, and continues to be, one of the central determining factors shaping Israeli society.

As a case in point, throughout the 20th century, the strengthening of the Palestinian national movement paralleled the crystallization of the Zionist movement. The British Mandate had delineated the contours of Palestine, and this enhanced the sense of national and territorial belonging among both national movements. While "the formation of 'Palestine' in the consciousness of the native population was not simply an automatic response to foreign encroachment and rule, or the uncritical absorption of European definitions of Palestine along biblical lines" (Doumani 1999, 16), Palestinian national consciousness co-arose with indigenous social, economic, and cultural networks; shared cultural practices; and historical memories during the 19th century. At the same time, for Palestinians, the territorial struggle with the Zionists enforced a sense of unity, common purpose, and Palestinian nationalism. Indeed, the 1988 Palestine Declaration of Independence states:

The Politics of Maps. Christine Leuenberger and Izhak Schnell, Oxford University Press (2020).

DOI: 10.1093/oso/9780190076238.001.0001

> Occupation, massacres and dispersion achieved no gain in the unabated Palestinian consciousness of self and political identity, as Palestinians went forward with their destiny, undeterred and unbowed. And from out of the long years of trial in ever mounting struggle, the Palestinian political identity emerged further consolidated and confirmed.[1]

Attempts by Israel to suppress the Palestinian national movement only enhanced the establishment of the PLO in 1965, and the Israeli occupation of the West Bank and Gaza in 1967 brought populations that were formerly under Jordanian and Egyptian occupation, respectively, together as they were now facing a shared occupier (Portugali 1991).

The Israeli-Palestinian conflict also manifested itself in the history of surveys and maps. From early attempts by the Hagannah to map Arab villages in preparation for the war in 1948, to their night-time raid on the offices of the Survey of Palestine to empty it of maps and material that were to be transferred to the Palestinian Arabs, to the Israeli government's suspension in 1968 of land registration in the West Bank and the closure of its land records to the public (Shehadeh 1997; United Nations Relief and Works Agency for Palestine Refugees in the Near East [UNRWA] 2012), surveying, mapping, and allocating land rights continued to be a profoundly political issue. Indeed, as we've noted earlier, to map is to know and to control. A struggle over land will inevitably also become a struggle over rights to survey and rights to map. How then have maps helped enhance Palestinians' sense of national belonging and how are maps implicated in the attempt to build a Palestinian state and resist the dominant geopolitics envisioned by the Israeli state? While a detailed treatise of the history of mapping practices in Palestine would go beyond the scope of this book we will sketch out some of the significant historical developments in Palestinian mapping practices, particularly during the period after the Oslo Accords when Palestinians' attempt to establish the State of Palestine went hand in hand with their effort to survey and map their territory, despite their limited territorial sovereignty and notwithstanding the troubled geopolitics of the region.

A Short History of Palestinian Mapping Practices

Historically, Palestinians had a fraught relationship with maps. Scientific mapping was a colonialist project. Maps were hardly used by locals to navigate the locales they knew best (J. C. Scott 1998); they were the tools of administrators,

[1] See the 1988 Palestinian Declaration of Independence: https://palestineun.org/about-palestine/.

government officials, and colonialists. Jewish immigrants, often trained and educated in Europe's educational institutions and universities, were immersed in the Western scientific paradigm, yet these knowledge practices were less widespread and less useful to the indigenous populations of Palestine at the time. During the British Mandate, as Jewish and British cartographers worked side by side making and designing maps, Palestinian Arabs were employed by the Survey of Palestine; however, they were often employed in lowly positions and were not involved in the actual production of maps. Furthermore, as mentioned earlier, before the dissolution of the British Mandate, the British had planned to transfer cartographic material relevant for the establishment of a Palestinian state, according to the UN Resolution 181, to Ramleh and then on to Cyprus; however, the Hagannah

> attacked by night the Survey [and] . . . spent 7 or so hours in which they were in full control of the building. They emptied [it] out [and] what was left was not of much quality to the other side. (Interview B 2011)

The Hagannah virtually stripped the Survey of Palestine, stole maps and other material, and left the Palestinian Arabs with "not much cartographic material." The material that finally did get transferred was of "very very little value" (Interview B 2011). As a result,

> within a few weeks after the British left, the Survey of Palestine became the Survey of Israel and was put in full operation and printed maps so that the Israeli side had full cartographic equipment acquired. The Arab side did have the Jordanian part . . . the Survey of Jordan which was run by the British at the time had some equipment and some maps of part of Palestine, but they were not as well equipped and . . . updated. (Interview B 2011)

Not only were Palestinians left with scant cartographic resources and expertise, but after the 1948 and 1967 wars the "overwhelming majority of Palestinian intellectuals found themselves outside Palestine" (Doumani 1999, 25). Palestinian experts in the diaspora often did not have access to local sources and there were few indigenous stable academic institutions in the Palestinian Territories. The geographical dispersion of Palestinian academics and the resource-poor academic infrastructure within "the country" affected Palestinian experts' ability not only to provide a Palestinian historiography but also to establish and develop a wide array of academic disciplines within Palestine. As a result, on the eve of the 1993 Oslo Accords agreement between Israel and the PLO, the Palestinians had few cartographic materials of value. They had acquired maps (often neither updated nor detailed) on the open market; some were acquired illegally (as Israel put severe restrictions on Palestinians' access to cartographic material) and others were produced in Beirut, Italy, and elsewhere. In fact, Palestinians' cartographic information was so scant that they came to the Israelis for advice. They had met some Israeli cartographers at international conferences, and those contacts led to Arafat's advisors visiting Tel Aviv University's geography department. According to an eminent Israeli cartographer, they "came here and they asked for some advice and help in the production of maps. They had also some contacts with the Survey of Israel"; they were seeking more accurate cartographic materials for both administrative and educational purposes (Interview B 2011). While Palestinians were frequently reliant on Israeli materials, what they could access was restricted. As a matter of fact, Israeli cartographers were prohibited from "providing them [Israeli Arabs (Palestinians within Israel)] with cartographic material and air photos which were restricted, not secret, [but] restricted" (such prohibitions thus affected Palestinians within Israel as well as within the West Bank). As a result, Israeli Arab students at the time were not allowed to sit in classes where this material was discussed (Interview B 2011). Such restrictions on Arab-Israeli students were in place until the mid-1990s, when they were slowly eased in the post-Oslo era. At that time, contacts between cartographic communities in Israel and the Palestinian Territories also increased, and Palestinian geographers and cartographers were increasingly trained in Israel and abroad and so were able to subsequently institutionalize these disciplines in Palestinian universities. Consequently, Palestinians' increased involvement in international academic networks enhanced their surveying and mapping capacity.

At the same time, a political shift transpired. The Palestinian drive toward national liberation by armed struggle was increasingly replaced by a reliance on international law and widely recognized UN resolutions to facilitate the state-building process. The language, tone, and appeals to legitimacy from the PLO's National Charter (formerly known as the covenant drafted in 1963) which was passed in 1968 differed quite drastically from the 1988 Palestinian Declaration of Independence. The PLO National Charter stated:[2]

> Article 1: Palestine is the homeland of the Arab Palestinian people. . . . Article 2: Palestine, with the boundaries it had during the British Mandate, is an indivisible territorial unit.

The Charter went on to emphasize that it was a Palestinian's national duty to be prepared "for the conflict and armed struggle as well as for the sacrifice of his property and life to restore his homeland, until the liberation is achieved."

[2] For the Palestinian National Charter 1968 see http://www.pac-usa.org/the_palestinian_charter.htm.

It also stresses the importance of "safeguarding Palestinian identity and developing a Palestinian consciousness of that identity" (Palestinian National Charter 1968), emphasizing the illegality of the establishment of the state of Israel and discrediting Jewish claims of a historical tie to the land of Palestine.

The 1988 Palestinian Declaration of Independence similarly maintains that "the Palestinian Arab people ensured for itself an everlasting union between itself, its land and its history," yet it also affirms that the 1947 UN Resolution 181, which terminated the British Mandate and partitioned Palestine in order to create a Palestinian and a Jewish state, provides legitimacy for Palestinian sovereignty within the Palestinian Arab-assigned territory.[3] The declaration affirms the PLO as its sole and legitimate representative and pronounces the Palestinians' "natural, historical and legal rights" to the land ascribed to them under the UN resolution. In doing so, it proclaimed the State of Palestine. While the declaration already recognized the inevitability of Palestine's territorial division, by 1993 Arafat changed the Charter and confirmed that those articles that deny Israel's right to exist were no longer operative or valid. This shift from a national struggle to liberate the whole of Palestine to the acceptance of a Palestine divided between a Palestinian and an Israeli state was increasingly reflected in Palestinian mapping practices.

And, indeed, during the 1990s Palestinian society embarked on a new era. In 1991, the peace process with Israel was instigated; in 1993, the signing of the Oslo Accords put Palestinian self-governance and final status negotiations with Israel on the agenda; and in 1996, the election of the Palestinian Legislative Council "created a new framework for Palestinian control over the planning process and development of the built-up environment" (MOPIC 1998, I). For the first time in history, Palestine was to be a state in the making, and Palestinians had a mandate to develop the territory under their jurisdiction. In fact, the State of Palestine was recognized as a UN non-member observer state in November 2012, and as of August 2018, 137 of the 193 UN member states and two non-member states had recognized it. However, the international recognition has not translated into sovereignty on the ground. While Palestinians' lack of full territorial sovereignty continues to constrain their ability to develop their jurisdictional space (as, for instance, planning for the presently still Israeli-controlled Zone C is hampered by their spatial planners' lack of access to the area and restrictions on their movements across it), they nevertheless attempt to plan for the eventuality of Palestinians gaining full sovereignty over the whole of the West Bank as specified in the Oslo Accords.

[3] For more information for the 1988 Palestinian Declaration of Independence see https://ecf.org.il/issues/issue/12. Retrieved Jan 27, 2020.

As we discussed in chapter 2, for the OECD, supporting nation-building, which entails forging a common national identity, and state-building, which includes building physical, economic, and political infrastructures and institutions, is central for international relations and stability. Just as surveying and mapping became some of the tools used to establish an Israeli state, they also support Palestinians' attempt to forge a State of Palestine. Consequently, Palestinian governmental and non-governmental institutions produced various maps for building the nation and building the state, respectively. In the following sections, we first discuss various generic Palestinian maps used for nation-building and then turn to the cartographic products used to support the Palestinian state-building process.

Maps for Palestinian Nation-Building

In the process of nation-building, "logo" maps that emphasize borders and help create the imagined community of a nation have always played a role in forging a sense of national identity (Anderson 1983; Jacob 2005; Jones 2008; Kosonen 2008; Paasi 1999, 2002; Radcliffe 2009, 2010; Weizman 2007; Wood and Fels 1992; Wood, Fels, and Krygier 2010). Logo maps of Palestine tend to delineate the boundaries of pre-1948 Mandate Palestine and symbolically depict Palestinian national belonging. Whereas in politically stable countries, maps that are used for nation-building converge with maps used for governance, in conflict regions, where territories and national histories are disputed, the boundaries depicted in nation-building maps may diverge from those represented in state-building maps. Due to the fragmentation of Palestinian-controlled territories, their transitional legal status, and the lack of clearly defined boundaries between Israel and the Palestinian National Authority, Palestinians tend to use the historical boundaries of Mandate Palestine as a "logo" map. Cartographic delineations of pre-1948 Palestine thus serve the project of nation-building as they help construct a citizenry that identifies with a bounded nation. As depicted in Figure 8.1, logo maps are used for official and popular events and they decorate posters, walls, art works, and handicrafts.[4]

While a depiction of historic Palestine with Jerusalem at its center and the Palestinian flag to its left can invoke a sense of nationalism, the symbol of a house key superimposed on historic Palestine (as depicted in Figure 8.2) evokes Palestinians' possession of their house keys to their pre-1948 homes (which they lost during the Nakba (between 1947 and 1949), including the 1948 war). The key symbolizes their right of return.

[4] See also collection at Palestinian Heritage Center, Bethlehem, http://giftshop.phc.ps/index.php/product/30-handmade-embroidered-wall-hanging.

Figure 8.1. Palestine Logo Map on Handicrafts. Photograph by Christine Leuenberger.

Similarly, the official symbols of the political parties Fatah (the largest political party within the PLO) in the West Bank and Hamas (an Islamic resistance movement that is associated with the Muslim Brotherhood) in Gaza reproduce the whole of Mandate Palestine. Hamas's commitment to the revolutionary struggle, which is outlined in its 1988 Hamas Charter,[5] is also reflected in its official symbol as represented in Figure 8.3.

Fatah's dedication to a revolutionary liberation struggle used to be reflected in its official symbol, too. However, the 2012 updated Fatah symbol, in Figure 8.4, reflects the political shift from liberating Palestine through armed struggle to making claims based on Palestinians' national, historic, and legal rights to the land. It depicts the whole of Mandate Palestine covered with a Palestinian keffiyeh (a traditional Arab headdress); it also contains the potent symbol of the house key, and it portrays the Al Aqsa mosque to emphasize Palestinian Arabs' religious link to the land.

[5] See Hamas Charter at: http://www.thejerusalemfund.org/www.thejerusalemfund.org/carryover/documents/charter.html?chocaid=397.

Figure 8.2. Palestine and the House Key. Photograph by Christine Leuenberger.

Logo maps are also frequently put into historical contexts, where they can do political advocacy work. Various Palestinian NGOs designed the map sequence in Figure 8.5, albeit with slight variations. The maps are juxtaposed to emphasize the Palestinians' loss of land to first the Zionist movement and then the State of Israel over time.

The initial map starts from the assumption that all land that was not sold in private transactions to Zionist settlers in Mandate Palestine constitutes Arab land (as indicated by the color green). It juxtaposes this scenario with the 1947 UN Partition Plan, a 1949 ceasefire lines map, and a map of post-Oslo Palestine (in which Palestinian sovereignty is limited to Areas A and B with Area C under Israeli control). The selection of these particular maps is significant as they all carry international legitimacy, while their visual language seemingly clearly and self-evidently reveals Israel's expansionist policies.[6]

[6] Israeli right-wing interest groups provide a parallel map-based story of land loss. One of their political leaflets, the Israeli *Non-Party Committee against Withdrawal*, presents a sequence of maps including maps of "The Israelite Kingdom of David and Solomon ca. 1000 B.C.E." "Palestine—The Jewish National Home of 1919" (which includes Palestine as well as Transjordan), and a map of

Figure 8.3. Hamas symbol. Available from http://hamas.ps/en/. Retrieved June 6, 2019.

Other ways of critiquing, challenging, and resisting the dominant geopolitics include cartographically reconstructing pre-1948 Arab topography, villages, and towns on maps of Mandate Palestine; such maps establish Palestinians' historical connection to Palestine and serve to politically legitimize their right of return to their families' places of origin. Atlases can also serve as a means to challenge dominant geopolitics. While atlases always have played a role in state-building projects, for a stateless people an atlas can become a powerful call for justice. In his atlas *The Return Journey,* Salmon Abu-Sitta (2007) retraces the Arab villages and towns before the establishment of Israel. He points out that "Palestine has been

"Israel—Armistice Lines 1949–1967" (which differentiates Israel from the West Bank and Gaza). The selection of different territorial configurations at vastly different time periods enables both Israelis and Palestinians to construct a cartographic story of land loss in different ways. Others juxtapose Israeli-controlled land in 1967 to territories controlled in 2001, as exemplified in *What Israel Has Given Away—for Peace*, and hereby again tell a story of territorial loss. Available at http://thinkingoutsidetheblog.blogspot.com/2011/06/44-years-later-israels-land-giveaway.html. Retrieved June 2, 2019.

Figure 8.4. Fatah's Official Logo since 2012.
Source: http://www.timesofisrael.com/new-fatah-logo-eliminates-israel/. Retrieved June 6, 2019.

Figure 8.5. Palestinian Loss of Land 1946 to 2010.

subject to an unparalleled massive displacement, dispossession, and replacement of its natural population in 1948" (Abu-Sitta 2007, 242). The maps and narratives not only reconstruct pre-1948 Palestine but also serve as a political advocacy tool. By pointing to Palestinian Arabs' dispossession of their land, the destruction of their property, and their diaspora status, he calls for their right of return. He points out that

> the majority of refugees around the world have returned to their homes after the cessation of hostilities. . . . But not in Palestine. Israel denied re-entry and return of Palestinians to their homes. . . . The Right to Return is a basic principle of human rights. Article 13(2) of the Universal Declaration of Human Rights states, "*Everyone has the right to leave any country, including his own, and return to his country.*" For Palestinians, the right to return is enshrined in the famous UN resolution 194, which is affirmed by the international body more than 135 times. (Abu-Sitta 2007, 232)

By mapping the Palestine of 1948 and listing all existing and depopulated villages before and in 1948, the map superimposes a past reality onto the current geopolitical territorial arrangements. By doing so, the atlas appeals to history to claim the land for Palestinian Arabs while at the same time emphasizing Israel's infringement of human rights as well as highlighting UN resolution 194 (which the UN passed near the end of the 1948 war delineating principles for reaching a final settlement and returning Palestinian refugees to their homes). In his map "Palestine 1948," in Figure 8.6, Abu-Sitta states that it belongs to the principles of justice that:

> Denial of the Right of Return is a perpetuation of Ethnic Cleaning which is a War Crime. Ethnic Cleaning must be reversed. Legally, the Return can ONLY be implemented to the refugee's original home and land in 1948, not anywhere else, not even in the future state of Palestine. There is no legal or moral justification for bartering a Palestinian State which is a political act of sovereignty, with the Right of Return, which is an Inalienable Right.

Abu-Sitta not only calls for the return of Palestinians to Palestine but also attempts to prove the feasibility of their returning with maps and statistics. Such cartographic reconstructions of history challenge the current geopolitics in the region. Accordingly, not only the Israeli state stands accused of war crimes but the post-Oslo Palestinian state-making enterprise is also presented as contrary to fundamental human rights as it would not enable diaspora refugees to return to their original home. Such maps therefore became powerful tools to advocate

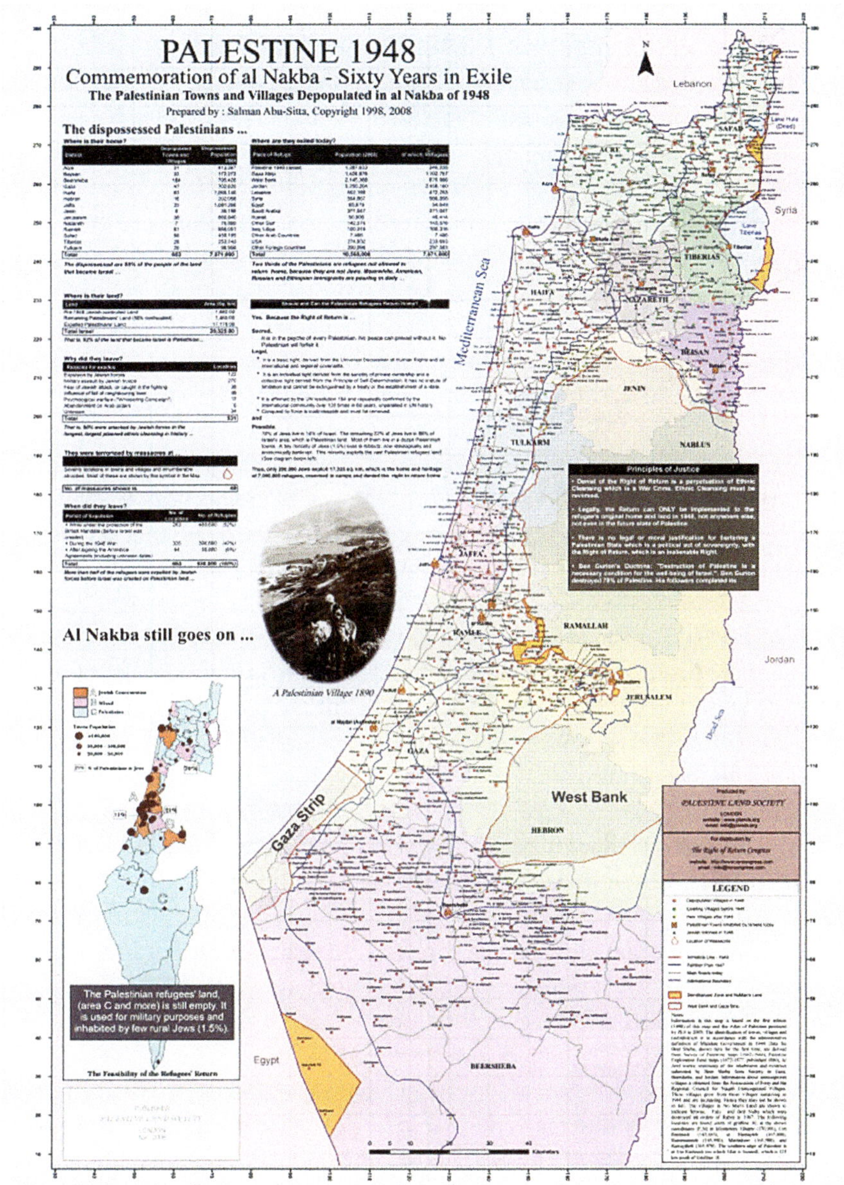

Figure 8.6. Palestine 1948.

Source: Cornell University: Artstor.

particular political positions and to provide a visualization of an alternative cartographic reality.

Besides such well-resourced projects, which are supported by diaspora communities abroad, after 1948, Palestinians also engaged in various "bottom-up" mapping practices that served to connect their people to the land, foster a sense of national belonging, and forge their mental maps of a Palestinian geography as a way of recreating "the lost homeland" (R. Davis 2007, 53). During the 1980s and 1990s, some Palestinian communities instigated the "village memorial book" project that collects information about village life in historic Palestine such as maps, land records, geologies, photographs, and stories; it also provides a "shared past" that presents a nationalist narrative about life before and after 1948; and it turns the "Palestinian peasant" into a "national signifier" that helps "create a nation with a unified culture and sense of authenticity" (R. Davis 2007, 70). Although around 40% of the Palestinian population in 1948 lived in cities, it is peasant life that is emphasized, which also become a way to claim the land. By preserving and glorifying peasant life, the maps and narratives in the memorial books prove the close connections the villagers have to the land, which in turn becomes a national discourse that serves as a tool for political advocacy.

> Given the Palestinian-Israeli struggle over land, the very act of creating the map of the past . . . claims an authority to know—by listing names of places, people show the knowledge of that place—and to imprint their presence on the land through the authority of knowledge. The authority serves to maintain Palestinians' ties to pre-1948 Palestinian land: by showing their intimate and familiar relationship to as well as their former dependence on the land, the maps help individuals continue to define themselves as Palestinians and as belonging to a particular village. (R. Davis 2007, 60)

As it was mostly colonialists and occupiers who wrote Palestine's history, such bottom-up attempts reconstruct a shared Palestinian past that has been largely neglected (Doumani 1999), and they provide an informal archive of local knowledge.

The maps discussed so far, ranging from logo maps to village memorial books, represent nation-building maps that serve to legitimize the nation, its territory, and its Palestinian descendants. Such maps, however, don't represent the territoriality of the state and its more limited definition of the right to citizenship. Yet building nation-states entails not only the construction of an imaginary community of a nation but requires also cognitive and material resources for the engineering of a state's infrastructure. In the following we will focus on Palestinian maps that are produced for state-building.

Maps for Building the Palestinian State

As we have seen, surveys and maps have always been crucial for state-building. They are fundamental for allocating land rights, developing resources, constructing infrastructure, and more generally, governing a territory. With the Palestinian state-building project under way, various governmental and non-governmental institutions, ranging from the Palestinian Central Bureau of Statistics to various Palestinian Authority (PA) ministries and the Applied Research Institute Jerusalem (ARIJ), among others, produced maps for the purpose of building the state.[7] Such maps differ from nation-building maps in that they delineate the West Bank and the Gaza Strip as separate from Israel in accord with international law and internationally recognized territorial agreements, such as the 1993 Oslo Accords. One of Palestine's flagship research institutes, ARIJ,[8] has produced an array of maps, including the "Geopolitical Map of the West Bank," in Figure 8.7, which portrays territorial delineations of Israel/Palestine according to international law.

To be sure, in the bid for fully recognized statehood, a people's strict reliance on internationally recognized territorial divisions is crucial to garner legitimacy and credibility for the eventuality of gaining territorial sovereignty and independence (Leuenberger 2012). This is particularly important in a resource-poor environment such as Palestine where mapping projects are frequently funded by international governmental and non-governmental development agencies and international donors. To follow internationally legitimized precepts is therefore not a choice but an obligation. Indeed, one Palestinian interviewee pointed out that if Israel was not named in the maps produced, funding would dry up instantaneously. As international funding as well as positive media coverage is conditional upon a people's reliance on transnational laws and treaties, incorporation of and reliance on international laws and treatises are "some of the only tools available to the struggle for rights of the disenfranchised" (Merry 2006, 49) and can become a form of political advocacy.

ARIJ's "Palestine Road Map" as reproduced in Figure 8.8, shows the intermixing of international treaties with local understandings of realities on the ground. This map clearly delineates the Green Line and the adjacent, yet diverging "Segregation Wall" (depicted as thick red lines with white sections), which in a 2004 advisory opinion the International Court of Justice declared to be illegal under international law.[9] The Green Line doesn't just assert

[7] For other state-building maps, see Palestinian Central Bureau of Statistics' *Statistical Atlas of Palestine*, http://atlas.pcbs.gov.ps/atlas/pages/AboutAtlas-en.asp.

[8] For more information about ARIJ, see https://www.arij.org/.

[9] International Court of Justice, *Legal Consequences of the Construction of a Wall in the Occupied Palestinian Territory: Advisory Opinion* (July 9, 2004), http://www.icj-cij.org/docket/index.php?pr=71&p1=3&p2=1&case=131&p3=6. Retrieved June 11, 2009.

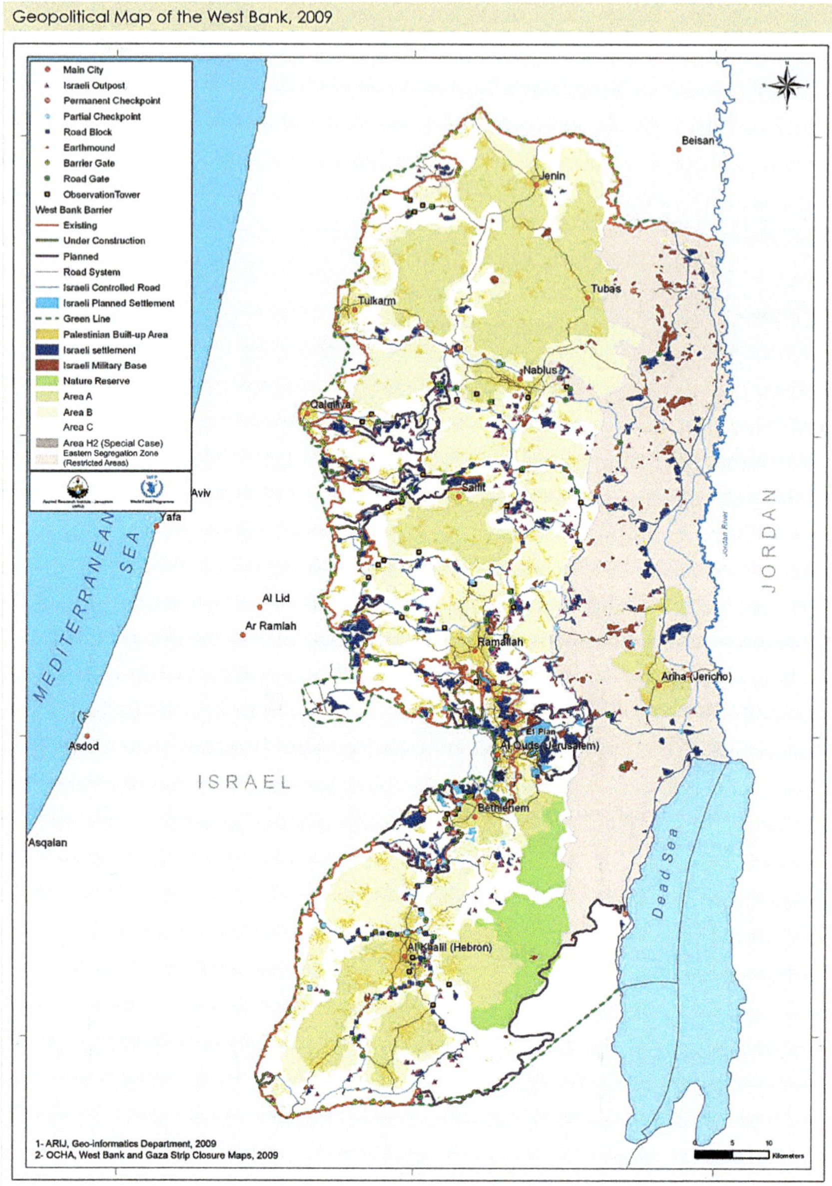

Figure 8.7. Applied Research Institute Jerusalem (ARIJ): Geopolitical Map of the West Bank. Courtesy of ARIJ.

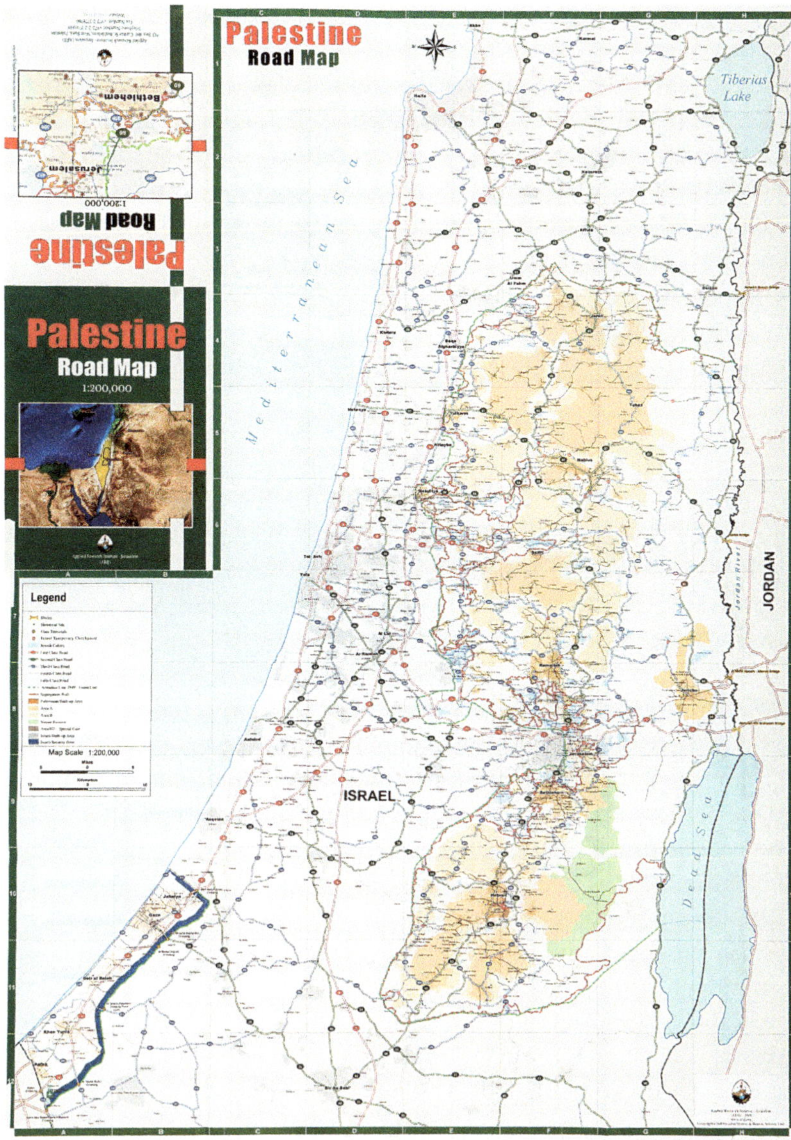

Figure 8.8. Palestine Road Map. Source: Cornell Artstor.

Palestinian land claims according to the 1949 armistice lines, but its visually prominent depiction of the wall also emphasizes its illegality as it deviates from the Green Line (portrayed as a broken green line) by veering, often deeply, into Palestinian territory. "The wall" is also represented as part of a larger system of exclusionary mechanisms, ranging from Israeli checkpoints to Israeli built-up areas, which impact the freedom of movement throughout the West Bank. Such a depiction serves not only as a source of information but also enhances Palestinians' ability to lobby for their territorial claims and to emphasize the impact of the Israeli occupation. By translating international laws and treatises cartographically into maps, these narratives not only allow for protest and resistance to the dominant hegemonic geopolitics, but they also provide a platform for combining local understandings on the ground with transnational notions of international law.

Despite its integration of legal treatises into the designing of the map, on the ground, just carrying such a map at an Israeli checkpoint can become a point of contention. As the map's producer pointed out in an interview,

> You know I put a copy of this road map in my car when I drive sometimes . . . and one time I came across a checkpoint and the Israeli army stopped us at the checkpoint and they saw the map and they got mad: "How come you have a map?" "What's wrong with having a map?" [I said], and [they said] "It says Palestine! Where did you get it?" "I did it. It's me." He couldn't believe it. So, he tried to call—as if he'd found—you know—a terrorist. And somebody higher up stopped this non-sense. (Interview JI 2011)

Thus, while the appropriation of international law may serve as an advocacy tool as it is perceived to be neutral and objective to some, others do not perceive international law as a neutral arbitrator, and its intermingling with local and often controversial understandings and naming practices can nevertheless fuel conflict over cartographic representations of what's on the ground (Leuenberger 2013).

ARJI also produces an *Atlas of Palestine*, which is the first comprehensive Palestinian-produced atlas of Palestine. It exemplifies the Palestinian political and cartographic shift toward striving for an internationally legitimized two-state solution. Indeed, in its foreword, Ahmad Quraai, speaker of the Palestinian Legislative Council, points out: "This Atlas is another building block towards Palestinian statehood" (ARIJ 2002, 1). Consequently,

> The occupied Palestinian territory (oPt) is divided between two physically separated areas known as the "West Bank" (including East Jerusalem) and the "Gaza Strip" with a total area of 5,661 sq km and 362 sq km, respectively. (ARIJ 2010, 2)

The various thematic and historical maps in the atlas clearly differentiate the West Bank and Gaza from Israel. International law, UN resolutions, and the Oslo Accord are cited as the stepping stones toward legitimate Palestinian statehood.

In conjunction with an appeal to such internationally recognized legal and territorial agreements, the atlas also presents graphs depicting "Five thousand years of Palestinian history," (ARIJ 2002, 7) which trace the history of the Palestinian people and their experience of occupation by a series of nations. Its combination of visual and textual narratives establishes the legitimacy of Palestinian land claims and their historical connection to the territory.

At the same time, according to an Israeli cartographer, the maps and narratives "give their view" (Interview B 2010) and presents the "Arab version" of the boundary delineations, as it includes the no-man's land around Latrun on the Palestinian side and shows the Green Line cutting through Jerusalem. As maps have always been representative of the perspective of a particular nation-state, taking that perspective is also indicative of the indigenous information-gathering and mapping efforts under way. As the atlas's editor points out, however, the "information-gathering pertaining to Palestine has endured severe restrictions and limitations" (ARIJ 2002, 3). He emphasizes the detrimental effects of the Israeli occupation on territorial development and planning, which have been particularly harmful given that as Ahmad Quraai points out, "Palestinians are waiting for the opportunity to build a modern Palestine" (ARIJ 2002, 2). The ability to delineate the territorial borders of what is expected to constitute a future Palestinian state is vital for the purposes of surveying, mapping, and spatial planning. Yet despite the importance of creating maps for state-building after the Oslo Accords, developments under the PA reveal the difficulty of spatial planning and map-making in a resource-poor, politically fragmented, and unstable context. What then are some of the social, institutional, and political ecologies of spatial planning and map-making for Palestinian state-building?

Planning for Palestine

Since the 1967 war between Israel and its neighboring states, the Israeli civil administration and the Israeli military have been in charge of administrative and legislative issues, spatial planning, and infrastructure development in the West Bank and the Gaza Strip. When the PNA gained some territorial jurisdiction following the 1993 Oslo Accords, Palestinians lacked the spatial data and the physical infrastructure necessary for planning and future development. The PNA, with assistance from the Norwegian Agency for Development Cooperation (Norad), initiated the Physical Planning and Institution Building Project (PPIB). Its goals were to build up the institutional infrastructure to foster physical planning; develop legislative and regulatory procedures (including policies, guidelines, and standards); provide training in spatial planning; enhance data sharing among stakeholders; and produce plans for development that span regional and national governing bodies (Abdullah 2001; MOPIC 1998; Scanteam and ARIJ 2009).

Palestinian spatial planners, however, continued to face numerous obstacles when planning for Palestine's national territory (El-Atrash 2016; El-Atrash, Salem, and Isaac 2009; Leuenberger and El-Atrash 2015; UN-Habitat 2015; UN Press Release 2014). For instance, the PNA inherited often contradictory, non-standardized, and incompatible legislation concerning land allocation and registration from previous regimes, including the Ottoman, British, Jordanian, Egyptian, and Israeli administrations.[10] Moreover, as the PNA has never been unified, it lacks not only a coherent political and administrative infrastructure but also spatial plans that span the regional and national levels. Despite the aim of increasingly coordinating "the non-governmental sector, institutions of local government, and the Palestinian Authority" (PA 2008, 17), such coordination is still lacking. Instead, the relationship between the government and NGOs varies between "cooperation, competition and outright conflict" (PA 2008, 33). Consequently, not only are planning efforts fragmented, but there are also unclear and overlapping mandates across different governmental and non-governmental institutions (Salameh 2008; Scanteam and ARIJ 2009). This fragmentation and duplication of efforts is magnified by different agencies' collecting data but failing to archive, share, disseminate, and update it. This is despite the fact that reliable, standardized, and up-to-date information is crucial for both spatial planning and peace negotiations with Israel (MOPIC 1998).[11] More generally, scholars have long argued that shared and legally enforceable standards, codes, and classificatory systems are crucial to state-building and effective governance (Bowker and Star 1999).

Additionally, the precarious geopolitics of the region hampers spatial planning. The territorial fragmentation of the Palestinian Territories into two separate areas, the Gaza Strip and the West Bank, as well as the latter's division into three zones, impedes planning projects and hinders the equitable and efficient distribution of infrastructure and services:

> Until now, the geopolitical situation of the [oPt] is uncertain. . . . What would be the border of the Palestinian state . . . what kind of sovereignty will Palestinians

[10] As discussed earlier, during the Ottoman and British Mandate periods, various individuals, organizations, and villages produced a range of non-standardized maps, which impeded the allocation of land ownership. Although the aim of the Survey of Palestine was to perform a full-scale cadastral survey, by 1948 only about 20% of the territory had been surveyed. The resulting uncertainties continue to inflame disputes over land ownership between Jewish settlers, the Israeli government, and Palestinians (Gavish and Kark 1993).

[11] According to Scanteam and ARIJ (2009), PPIB was criticized for collecting data solely for its own planning purposes without disseminating and sharing the collected information. As a result, PCBS [the Palestinian Central Bureau of Statistics] continues to use "paper maps" rather than digitized maps, and the Palestinian negotiating team does not have access to an agreed upon map of a future Palestinian state that could be used in final peace negotiations (Scanteam and ARIJ 2009).

> have . . . is definitely uncertain. Under these circumstances, one questions the value of national or regional planning, but at the same time, it is not wise to wait and do nothing until a peace agreement is signed. (Scanteam and ARIJ 2009, 19)

While currently the PA administers only Zones A and B, according to the Oslo Accords, its jurisdiction is to include Zone C in the future. Planners therefore adopt the assumption that the pre-1967 West Bank border delimits the planning area. But while they prioritize planning for Zones A and B, they also need to plan for Zone C, despite lacking access to these still Israeli-controlled areas.

> Because as a consequence of the Oslo Accord the Israelis will withdraw from the whole West Bank—so we are establishing ourselves technical-wise to rule the area so we need to prepare the plans. . . . [T]he agenda was—we would have the West Bank and Gaza and there will be an influx of returnees and there is a developmental vision for the Palestinian state. (Interview Z 2011)

Moreover, planning for East Jerusalem as the future capital of Palestine is particularly fraught given various Israeli government policies that are geared toward annexing territory under its jurisdiction while using planning as way to do so (Habiballah 2016; Nolte and Yacobi 2015; Leuenberger 2016a). Other political constraints and instabilities—such as the need for Palestinian legislation to be approved by Israeli governing bodies; the inability to implement national planning projects until there is a clear outcome from future peace negotiations; and the constant possibility of political instability, such as the Intifada and Israeli checkpoint closures—also impede planning efforts. As we have seen, there are many factors, including fraught political histories, institutional fragmentation, territorial discontinuity, and numerous geopolitical uncertainties, that negatively affect spatial planning efforts. How, then, does this political and institutional landscape affect map-making for the purpose of state-building in the PA?

Making Maps for Palestine

Geographic information and spatial planning have always been critical for utilizing natural resources and land, increasing economic growth, building infrastructure, expanding social services, raising standards of living, and improving political decision making. Reliable spatial information is, however, particularly decisive in territorial disputes. After Palestinians found, that in early negotiations with Israel, that their data and information were "scant when compared to what the Israeli negotiators had at their fingertips," (Scanteam and ARIJ 2009, 5) and that they often had "no maps of their own" (Scanteam and ARIJ 2009, 14), the aim was to build up cartographic expertise for future peace negotiations.[12] As one interviewee, a researcher at a Palestinian NGO, concerned with spatial planning and development, pointed out:

> We should have been prepared when they started with the peace process [but] apparently the Palestinian side went to the negotiations with virtually no maps; they were using Israeli maps [and] they were using other people's maps. But no maps were used by the Palestinians—[that were done] from the Palestinian side. This had its consequences later on as we see today. (Interview J 2011)

Given that the lack of maps from a Palestinian perspective had contributed to failed peace negotiations and detrimental consequences for Palestinians and their land claims, he pointed out that evidence-based proposals are also important for political advocacy in international institutions:

> We are trying to . . . put more facts into our research or in our publications, in our presentations to the world—[to show] how to see the conflict on a factual basis from the Palestinian perspective—because—apparently we have not been doing so good in the international arena when it comes to showing and exhibiting the facts on the ground. . . . [T]he first questions would be "Where is your source, what are your facts?" (Interview J 2011)

Therefore, when arguing for the Palestinian position with international bodies, such as the United Nations, evidence-based arguments rather than emotive appeals are seen as crucially important by Palestinian protagonists in order to further Palestinian political interests. As part of such efforts to provide more evidence-based policy proposals, surveys and maps have become an important tool in their toolbox. Consequently, due to efforts by various governmental and non-governmental organizations,

> from about 1998 onward there were maps and data on Palestine from internal sources. These maps were of utmost importance in the assessments of Israeli proposals, and their intended use was to develop the Palestinian position on such diverse subjects as territory, boundaries, land annexation and swaps, Israeli settlements, water, resettlement of Palestinian refugees, security, and the sharing of Jerusalem. (Scanteam and ARIJ 2009, 14)

[12] For instance, when in 1994 the Israeli government agreed to withdraw from Jericho and most of Gaza, the maps on which the negotiation was based were produced by Israel. "In fact, the Green Line was not outlined in the Gaza/Jericho agreement which raised a number of questions among Palestinian negotiators and necessitated better preparedness in the future" (Scanteam and ARIJ 2009, 6).

While collecting spatial data is vital to improve governance and to prepare for future peace negotiations with Israel, various institutional and geopolitical factors continue to impede the mapping of the Palestinian Territories.

First, a lack of human, technical, material, and financial resources affects map-making for state-making purposes. For instance, many training modules in surveying and data collection as well as cartographic projects that aim to produce and digitize paper maps and land-registration documents are supported by such international donors as the World Bank and USAID (World Bank 2016). As these kinds of projects are funded on a project-by-project basis, the feasibility of building up, maintaining, and updating a national geospatial infrastructure and database over the long term is uncertain. Second, the precarious geopolitical reality on the ground also affects the ability of planners and cartographers to survey and map territories. For instance, Israeli-imposed security restrictions hamper Palestinians' access to aerial photos, satellite images, and surveying data; they may be provided with only low-resolution aerial photos on which politically sensitive areas (such as Jewish settlements in the West Bank, Israeli military infrastructure, and aquifers) are eliminated, as seen in Figure 8.9.

Figure 8.9. Aerial photos with Israeli settlements and military infrastructures erased. Photo by Christine Leuenberger.

As an interviewee remarked,

> For aerial photos . . . that include settlements or military bases . . . they will remove them. . . . [W]ith photoshop they removed the settlement and put something vacant as though there was nothing and then we compared it with google earth. (Interview JI 2011)

Israeli security restrictions also impact the ability to survey the territories. For instance, according to a World Bank report (World Bank 2009), because aerial photographs at a suitable scale were not available, a West Bank survey could not be completed. Moreover, aerial photographs of the Ramallah Municipality (an Area A region under Palestinian control) could not be obtained due to "restrictions imposed by the Government of Israel on flights over the West Bank" (World Bank 2009, 15). Internal political instabilities, too, such as the takeover of Hamas in the Gaza Strip, can halt surveying, mapping, and planning projects. Third, Palestinian scholars and international development agencies have pointed out that "the governments in the Palestinian Territory are fragile, and in view of the fact that a Palestinian State has so far not been established, institutional sustainability must be assumed to remain elusive" (Scanteam and ARIJ 2009, 23). For instance, since 1995, the Palestinian Ministry of Planning (MOPIC) has experienced major transformations. Over a six-year period, six different ministers were in charge, and their planning priorities continued to change along with frequent staff turnovers. Also, although various governmental and non-governmental agencies agreed on new land reforms and land laws, in 2009, "the passage of these laws by the Palestinian Legislative Council has not taken place as the Council has not been in session since June 2007" (World Bank 2009, 14). Moreover, even if these laws were passed, no institution could enforce compliance. Fragile governing institutions thus hamper the implementation of systematic, sustainable, and long-term strategies for development and planning.

Such tenuous and fragmented institutional landscapes also contribute to the lack of standardized geographical data:

> Palestine suffers due to the non-existence of a national mapping agency responsible for the creation of base maps. This situation had resulted in a lack of data, many uncoordinated activities, and lack of institutions that can play a relevant role in the process of the development of society. (Abdullah 2001, s.5.1.3)

Because no institution has the political mandate to produce authoritative national maps, there is still no "unified base map for Palestine that is accepted by all institutions" (Scanteam and ARIJ 2009, 17).[13] Different governmental

[13] Scanteam and ARIJ (2009) point out that even if one institution were to gain the political mandate to be the PA's official cartographic unit, an institutionally weak legal system could not enforce its political and legal mandate to produce national maps.

and non-governmental units have their own cartographic departments; each produces its own maps, none of which adhere to a unified system of codes and standards; and many maps are duplicated because of a lack of data sharing between different mapping units. As a result, there is a range of non-standardized maps across diverse governmental and non-governmental institutions. Stakeholders maintain that such weak institutional infrastructures with no clear political mandate produce "a poor climate for legislative work and joint policy, regulatory, procedural and standards development" (Scanteam and ARIJ 2009, 15). Indeed, the absence of both a national geospatial database and uniform cartographic standards impedes all kinds of state-building projects, including the allocation of land ownership, the development of land resources, the production of national spatial plans, and the sharing of data across institutions.[14]

Weak state institutions, established in a resource-poor environment and characterized by a lack of cooperation and standardization across different institutional settings, are common to most developing countries. Palestinian institutions are unique, however, in that they also operate within a precarious geopolitical reality that includes a lack of territorial sovereignty and the absence of a peace agreement with Israel. The combination of these factors not only hampers legislative work, policymaking, and spatial planning but can also have a negative impact on the international legitimacy of the Palestinian state-building process. Indeed, the establishment of an authoritative mapping agency with the political mandate and the resources to enforce cartographic standardization and regulate the delineation of national boundaries across nation-building and state-building maps would not only assure the international community of Palestinians' intention to coexist peacefully with Israel but would also increase the legitimacy, credibility, and visibility of the Palestinian state-making process.

Maps in the Crossfire

Israeli critics, from online bloggers to the Israeli government, have produced an array of documents to show that Palestinians produce maps that do not recognize Israel. On the web, it is all too easy to find materials that seemingly support this point of view. There are maps of the whole of Palestine that omit Israel, as well as websites that support Palestinians' exclusive right to their homeland.[15]

[14] Abdullah (2001) calls for a National Spatial Data Infrastructure, which would act as a national data clearinghouse to coordinate national and cross-national mapping activities; set standards and policies; provide data storage to minimize data redundancy and duplication; raise political awareness of the importance of spatial data for national development; and maximize data integration and dissemination.

[15] Pro-Israeli websites critique Palestinians maps (ranging from maps in educational textbooks to etchings and governmentally produced maps) for "instilling fundamentally negative messages relating to the Israeli-Palestinian conflict. These include ignoring the existence of the State of Israel,

Such online documents enable critics to easily construct a narrative to support Israeli fears of Palestinians' lack of recognition of current geopolitical realities. Underlying this political point-scoring, however, is a deeper story about the neglect and disregard of the social, political, and cultural context of mapping practices within the Palestinian Territories. When focusing on the context of map production, we need to ask, for example, Who produces maps, for what purposes, and for what audiences? How do existing institutional and political conditions enhance or hinder certain mapping practices? Also, how standardized are mapping practices across different governmental and nongovernmental institutions? As we showed here, Palestinian institutions, political interest groups, and activists employ various generic mapping genres for the purpose of building the Palestinian nation-state. The analysis presented here doesn't provide an exhaustive account of their diverse mapping practices but focuses exclusively on two specific genres of maps that pertain to Palestinian geopolitics: maps produced by various actors, including artists, political activists, digital designers, and bloggers for the purpose of nation-building, as well as maps produced by the PNA and affiliated NGOs for the purpose of state-building.

The urgency with which Palestinian stakeholders survey and map the territory under their control speaks to the historic importance of maps in the service of state-building. While literature in critical geopolitics has increasingly advocated for counter-mapping, citizen mapping projects, and anti-geopolitical alternatives over officially endorsed and hegemonic cartographic projects, states-in-the-making, such as Palestine, reaffirm the importance of standardized mapping practices in the service of state-building (Koopman 2011; Megoran 2011). Especially in situations of conflict, officially sanctioned and legitimized cartographic practices can become a crucial tool for establishing and settling land claims and thereby also help establish citizens' rights. In a world dominated by a nation-state system that draws sharp borders between states and allocates citizens' rights based on national belonging, weak states and stateless citizens are severely disadvantaged. Striving toward legitimate statehood is not only crucial for the international economic and political system (OECD 2009) but is also a means to ascertain the rights of otherwise stateless citizens in a global order. Delineating territorial sovereignty and borders is a pragmatic step toward a more equitable

and denying the bond between the Jewish people and the Holy Land; the obligation to fulfill the Palestinian 'right of return'; the continuation of the 'armed struggle' for the 'liberation' of all of 'Palestine,' and perpetuating hatred of the State of Israel"; see http://www.jewishvirtuallibrary.org/jsource/History/palmatoc1.html (p.1). For how Israeli-produced maps also frequently de-signify and eliminate Palestinian topography, culture, and history and "judaize" and hebraize the landscape, see Falah (1996) and Leuenberger and Schnell (2010).

distribution of land and resources in a global political system in which citizens' rights and obligations have become tightly linked to 19th-century notions of nation-states.[16] This is why geo-informatics, including surveys and maps, has always been crucial for enhancing development, using natural resources, and improving governance.

[16] However, the formation of two states in Israel/Palestine does not preclude other possible region-wide political reconfigurations in the future, such as a binational federation between Palestine/Israel within the context of a Middle East Union. See Jeff Halper, "A Middle East Union: A Two-Stage Solution to the Israeli-Palestinian Conflict" (May 2004) http://www.mediate.com/articles/halperj1.cfm, retrieved December 18, 2012, or the "Two States, One Homeland" Initiative—see http://www.alandforall.org/en, retrieved June 3, 2019.

9
Conclusion

We live in a visual era in which images, graphs, and maps are ever more important in advertisements, in political and academic debates, and in policy circles. The power of visual rhetoric makes reflecting on its substance, functions, mechanisms, meanings, and impacts all the more valuable. In this book we have focused on maps and how they came to be and are used in public debates over territories in Israel/Palestine. In an era marked by calls for new ways of making and disseminating knowledge claims that go beyond the academic silos of the past, we mapped out a range of theories and methods that provide effective tools to analyze maps as part of contemporary culture (DEA 2012; Senge and Scharmer 2001). The ability to combine insights from across the social sciences enables us to understand the visual rhetoric of maps within the social, discursive, and political contexts of their production and use. As maps have become ever more part of public discourse, it is paramount to reflect on their content, function, and impact.

Israel/Palestine provides a particularly rich laboratory to study mapping as its territories and boundaries continue to be controversial, providing mapmakers fodder for cartographic warfare. In such areas of territorial conflict, it becomes clear that we cannot gain an "objective," bird's-eye view of a territory and map it accordingly. As political positions harden, it is all the more important to appreciate mapping practices and territorial claims-making from the inside so as to better understand nationally based discourses about land, territories, borders, and politics. In an era of a seemingly trans-national and globalized world it seems paradoxical to observe instead the reassertion of the power of the nation-state. Yet, scholars have noted that since the end of the Cold War and the collapse of the Berlin Wall, concerns about strengthening and implementing borders along with notions of national territoriality and sovereignty have reasserted themselves within the international political system (Popescu 2012; Vallet 2014). Focusing on nationally based cartographic discourses not only provides us with insights into the complexity, fissures, and frictions within internal political debates but also reveals the persistent power of the nation-state as a framework for forging identities, citizens, and alliances. As we have seen in both Israel and Palestine, governmental and non-governmental organizations, interest groups, and political protagonists produce a range of recurrent genres of maps for the purpose of nation-building

The Politics of Maps. Christine Leuenberger and Izhak Schnell, Oxford University Press (2020).

DOI: 10.1093/oso/9780190076238.001.0001

and for state-building. They have different levels of legitimacy, reveal different contents, and serve various functions. While many map-makers drew on cartographic conventions to legitimate their claims, some appealed to emotions such as fear of the "other," and others to human rights and a people's historical rootedness in the land. Their respective legitimacy, content, and function depend on the map's producers. But it is equally important to understand the institutional and political conditions that enhance or hinder certain mapping practices. This is crucial for gaining an understanding of the kaleidoscope of mapping practices present in Israel/Palestine.

Historically, "maps blossom in the springtime of the state" (Wood, Fels, and Krygier 2010, 15), but they also provide a lens through which to understand the making of nation-states. Since their rise in the 19th century, nation-states have become the predominant way to organize a global political system. As a result, striving toward legitimate statehood has become crucial for international relations and the stability of this system. Independent, sovereign, stable, effective, legitimate, and resilient states are now its bedrock, and its stability depends on member states' ability to manage scarce resources; secure, protect, and provide services to their citizens; enable orderly economic exchange; conform to certain standards of legal conduct; and assure political transparency, accountability, and legitimacy. Yet the formation of states has historically also been associated with violence, oppression, and repression, as exemplified by the long history of colonization and state-building in Africa and the Americas, among others. The establishment of Israel and the concurrent dispossession of the Palestinian Arab population in the region thus reflects the 19th-century history of colonization and state-building. Once established, states also have their downsides. Many states suppress ethnic and religious identities, enforce taxation while providing minimum services to citizens, and fail to provide political representation, institutional accountability, and the rule of law (OECD 2009). Also, weak and failing states—from the Middle East to sub-Saharan Africa—are not only a problem for political scientists but they can also lead to regional and international instability, human rights abuses, ethnic and political violence, and genocide. At the same time, in a world dominated by a nation-state system (Agnew 2007a, 2007b) that allocates citizens' rights based on national belonging, weak states and stateless citizens are severely disadvantaged. Delineating territorial sovereignty and borders can ascertain the rights of otherwise stateless citizens in a global order. In the case of the Israeli state-building process we have seen how it empowered certain social groups while disenfranchising others. Yet in the Palestinian case, the lack of an independent state continues to negatively impact the Palestinian people, as they remain a stateless people with neither adequate legal representation nor territorial sovereignty for surveying, mapping, and developing their territory.

Developments in Israel/Palestine exemplify how and why surveys and maps became fundamental tools for delineating territories, parceling property, and claiming land. In conjunction with the cartographic tools for state building there were, however, other cultural, political, and social prerequisites that helped Israel, as a nation-state, come into being. To engineer the imagined community of a nation—they needed maps but also commonly shared narratives, ideas, rituals, and practices. The Israeli state-building process would not have been possible without the strength of the Zionist vision, Judaic beliefs, and social rituals that enhanced the hegemonic and cultural cohesiveness of the Yishuv and the "recruited elite" that helped the early state-making efforts blossom.

By examining pre-state and early state mapping in Israel we saw how maps can help forge a territorial imagination and build the imagined community of a nation. Jewish organizations before and after 1948, such as the Jewish National Fund, crucially facilitated the implementation of the Zionist project. In conjunction with shared cultural practices—including celebrating the Shabbat and practicing various religious rituals—that helped maintain and enforce a unique Jewish cultural identity, a European orientalism and a modernist ethos that the Jewish "recruited elite" shared with the British colonizers, and a hegemonic Zionist vision, JNF maps of Greater Israel helped transform a Jewish-religious identity into a national-territorial one and thereby secularized the Messianic mission to redeem the "Promised Land" and develop an allegedly desolate region into a modern nation-state (Gurevitz 2007; Schnell 1999). The Blue Box in particular, with its boundless territory that was open to be redeemed by the Jewish diaspora, provided a logo map of the land and helped establish the myth of the open, yet to be settled frontier that persists today in the form of current neo-Zionist attempts to define the oPts as the new frontier of the "Promised Land" (Kimmerling and Migdal 2003; Sheleg 2000). The Blue Box, together with other nation-building maps that were produced by various governmental and non-governmental organizations, helped to "judaize" the territory and build the nation.

However, states cannot only be imagined; they also have to be engineered, and surveys and maps are crucial tools in that process. Pre-state mapping efforts under the British Mandate established a sizable cartographic expertise in the region that fell into the hands of the Hagannah in 1948 during an infamous nightly raid on the Survey of Palestine. As a result, at the end of the 1948 war, the Survey of Palestine was turned into the Survey of Israel, enabling the new state to fulfill all the functions necessary for surveying and mapping in the service of building the state. The surveyors and map-makers saw themselves as the pioneers in the Zionist effort to establish a state, and the power of the law, science, and technology, along with a modernist ethos, helped transform Arab Land into Israel Land and established a new ethnic geography within Israel/Palestine (Yiftachel 2006). Cartography had become an integral part of transforming the land into a state territory, developing the country, attempting to define its borders, and defending the nation-state (Routledge 1996; Koopman 2011; Megoran 2011). However, as we have seen from the discussion of constructing Israel's borders—surveys and maps are crucial, yet the social networks of collaboration, trust, and a sense of common purpose in the end are also critical to be able to get the job done—whether this entails building a state or delineating the boundary with adjacent states.

Yet after the 1967 war Israel had more than tripled the territory under its control, which marked the beginning of the breakdown of a territorial consensus and resulted in a "war of maps." The hegemonic Zionist vision enforced by networks and mutual trust that—fueled on by a common purpose—helped build the Israel nation-state was slowly fragmenting. Various Israeli social and political interest groups started to design maps to put forth varied geopolitical visions and demarcate disputed boundaries in different ways. New GPS technologies, online cartographic software, and access to the Internet turned an expert's cartography into a people's cartography (Crampton and Krygier 2005). Yet it is here where we may also encounter the limits of the democratization of mapping practices, as maps produced by lay activists, who often lack socio-technical expertise and/or institutional legitimacy, may reach no more than a committed in-group (Bittner, Glasze, and Turk 2013; Bittner et al. 2011).

By the 1970s Israel had been surveyed and mapped, and 93% of the land had become state land. The project of transforming Arab Land into Israel Land had been successfully accomplished, yet the de facto annexation of the West Bank, Gaza, and the Golan Heights opened up deep rifts within the Zionist vision. At that time, the nation-building parties increasingly lost their hegemonic power and thereby their ability to shape the Zionist vision. Instead, various formerly marginalized segments of society moved increasingly toward the center stage of politics and public debate (Mautner 2008; Kimmerling 2001). Jewish citizens were increasingly pitted against Palestinian Arabs, the Ashkenazi (from European origin) against the Mizrachi Jews (from Arab and Muslim countries), the national-religious and the ultra-orthodox against the secular segments of society, and the post-Zionist against the neo-Zionists. Indeed, some commentators remarked that if Israelis did not have a common enemy in the Palestinians, the fragile seams that hold the tapestry of Israeli society together would have torn a long time ago. Accordingly, Israel became a sectarian society that has become increasingly fractured:

> based on national, ethnic, religious, cultural, political and socioeconomic factors. The conflicting interests between various sectors generate fierce pressure and tensions that undermine the very foundations of Israel's social integration, collective identity, and the legitimacy of its institutions. These processes

> can potentially put an end to the era of Zionist ideology, by replacing Zionism with alternative ideologies, either particularistic in nature (such as Jewish-religious) or more universal in nature (such as post-Zionism or anti-Zionism). (Yuchtman-Yaar and Shavit 2004, 345)

While even during the pre-state and early state efforts, a mosaic of divergent groups with competing interests characterized Israeli society, at the time, a general consensus about the Jewish national goals superceded sectarian and ideological rivalries, tensions and fissures (Rebhun and Waxman 2004).

The increasing fragmentation of this communality of purpose has been accompanied by an increasing shift of power toward the political right wing and the neo-Zionist camp (Yiftachel 1997; Yuchtman-Yaar and Shavit 2004). Consequently, the collectively shared socialist-Zionist vision from the early state period is becoming ever-more transformed into a religious Zionist vision. By 2015 some of Israel's most notable commentators declared the death of the two-state solution as the political and defense establishment as well as the media became increasingly dominated by a new elite whose commonly shared vision is based on ideas of the redemption of the "Promised Land," the superiority and legitimacy of the Jewish project to settle the land, and orientalist/racist beliefs about "others," particularly Palestinians (Levy 2015).

The fragmentation of Israeli society that commenced in the 1970s went hand in hand with the rise of neo-liberal economic arrangements, which strove to limit governmental involvement in the national economy and the rise of private enterprise (Kimmerling 2001; Mautner 2008). This shift also introduced a range of new mapping technologies that enabled a variety of actors, institutions, and organizations to put forth their particular territorial vision of the land (Radcliffe 2009). Bottom-up mapping projects flooded the media, the market, and the internet. The Survey of Israel started to provide an online website where NGOs and private stakeholders could download maps and map layers in order to design and construct their own maps. As a result, individuals, organizations, and institutions could all enter the map war. The war was over visual, textual, and spatial signifiers that could attach different names to the same sites; delineate spaces in order to either "judaize" or "arabize" them; and claim them as belonging to a particular imagined community.

The fragmentation of Israeli society was inversely reflective of Palestinian conditions too. While Palestinian stakeholders have attempted to build the Palestinian nation-state ever since the early 1990s, their efforts remain fragmented (Abdullah 2001). The Israeli state-making enterprise confirms that a sense of unity is crucial to achieve a common goal and fulfill a shared vision. Palestinians who are subject to external as well as internal political, economic, and sociocultural forces that enhance the fragmentation of their society and their politics (Jamal 2006)—including Israel's occupation policies and Palestinians' competition for positions of power—continue to struggle to forge a hegemonic vision of a future state that is shared among its people. At the same time, progress is difficult given the lack of natural and economic resources, social and cultural capital, and the lack of regional and transnational political support for efforts to establish a Palestinian state.

The urgency with which Palestinian stakeholders survey and map the territory in an effort to establish a Palestinian nation-state speaks to the historic importance of maps in the service of state-building. Various Palestinian governmental and non-governmental institutions are producing maps to build the nation and construct the state. While some focus solely on producing cartographic knowledge for the sake of constructing a state territory, others challenge the territorial visions of the Palestinian state-in-the making and also resist Israel's dominant geopolitics of the region. While literature in critical geopolitics has increasingly advocated counter-mapping, citizen-mapping projects, and anti-geopolitical alternatives over officially endorsed and hegemonic cartographic projects, both the making of the Israeli state as well as the attempt to establish a State of Palestine reaffirm the importance of geo-informatics in the service of state-building for settling land claims, establishing citizens' right, enhancing development, using natural resources, and improving governance. Maps continue to serve as crucial tools in building nation-states, but they also have become tools of resistance, as activists design maps that continue to hunt the visionaries of dominant geopolitics.

Bibliography

Abdullah, S. Abdullah. 2001. "An Approach towards the Development of National Geographic Information Strategy in Palestine." Master's Thesis, University of Nottingham.

Abu-Sitta, Salman H. 2007. *The Return Journey: A Guide to the Depopulated and Present Palestinian Towns and Villages and Holy Sites (in English, Arabic and Hebrew)*. London: Palestine Land Society.

Abu-Sitta, Salman H. 2010. *Atlas of Palestine 1917–1966*. London: Palestine Land Society.

Ackerman, Spencer. 2010. "Monitor Israel's West Bank Settlements on Your IPhone." *Wired*. https://www.wired.com/2010/09/monitor-israels-west-bank-settlements-on-your-iphone/.

Adler, Katya. 2010. "Archaeology and the Struggle for Jerusalem." *BBC News*. Jerusalem. http://news.bbc.co.uk/2/hi/middle_east/8500332.stm.

Adler, Ron, and Dov Gavish. 1999. *50 Years of Mapping Israel: 1948–1999*. Tel Aviv: Survey of Israel.

Agar, J. 1977. "Maps and Propaganda." *Bulletin of the Society of University Cartographers* 11: 1–15.

Agnew, John. 2007a. "Know-Where: Geographies of Knowledge of World Politics." *International Political Sociology* 1 (2): 138–48.

Agnew, John. 2007b. "No Borders, No Nations: Making Greece in Macedonia." *Annals of the Association of American Geographers* 97 (2): 398–422.

Alatout, Samer. 2008. "'States' of Scarcity: Water, Space, and Identity Politics in Israel, 1948–59.'" *Environment and Planning D: Society and Space* 26 (6): 959–82.

Alatout, Samer. 2009. "Bringing Abundance into Environmental Politics." *Social Studies of Science* 39 (3): 363–94.

Almog, Oz. 1997. *Ha-Tsabar: Deyokan/The Sabra: A Profile*. Tel Aviv: Am Oved.

Almog, Oz. 2000. *The Sabra: The Creation of the New Jew*. Translated by Haim Watzman. Berkeley: University of California Press.

Amara, Muhammad, and Sufian Kabaha. 1996. *Divided Identity: A Study of Political Division and Social Reflections in a Split Village*. Giv'at Haviva: Institute for Peace.

Amiran, David, ed. 1985. *Atlas of Israel*. Tel Aviv: Survey of Israel.

Anderson, Benedict. 1983. *Imagined Communities: Reflections on the Origin and Spread of Nationalism*. London: Verso.

Andrews, Malcolm. 1999. *Landscape and Western Art*. Oxford: Oxford University Press.

Arieli, S. 2011. "The Geneva Initiative—Ex-Establishment Initiatives for the Bordering of Israel, a Critical Analysis." In *The Political Role of Maps in the Israeli-Palestinian Conflict*, edited by Izhak Schnell and Ephraim Lavie. Tel Aviv: Steinmetz Center for Peace, Tel Aviv University.

ARIJ. 2002. *Atlas of Palestine (The West Bank and Gaza)*. Jerusalem: Applied Research Institute–Jerusalem (ARIJ). https://www.arij.org/publications/books-atlases/72-2002/219-an-atlas-of-palestine-the-west-bank-and-gaza.html.

ARIJ. 2010. "Socio-Economic and Food Security Atlas in the Occupied Palestinian Territory." Jerusalem. https://www.arij.org/publications/books-atlases/80-2010/245-socio-economic-food-security.html.

Avineri, Shlomo, ed. 1971. *Israel and the Palestinians: Reflections on the Clash of Two National Movements*. New York: St. Martin's Press.

Ayoub, Nizar. 2004. *The Israeli High Court of Justice and the Palestinian Intifada: A Stamp of Approval for Israeli Violations in the Occupied Palestinian Territories*. Ramallah, West Bank: Al-Haq—Law in the Service of Man.

Azaryahu, Maoz. 1996. "The Power of Commemorative Street Names." *Environment and Planning D: Society and Space* 14 (3): 311–30.

Azaryahu, Maoz, and Arnon Golan. 2001. "(Re)Naming the Landscape: The Formation of the Hebrew Map of Israel, 1949–1960." *Journal of Historical Geography* 27 (2): 178–95.

Azaryahu, Maoz, and Rebecca Kook. 2002. "Mapping the Nation: Street Names and Arab-Palestinian Identity: Three Case Studies." *Nations and Nationalism* 8 (2): 195–213.

Azaryahu, Yosef. 1929. "Hebrew Education in Palestine." In *The Jubilee Volume of the Teachers' Union*, edited by David Kimche, 112–57.

B'Tselem. 2004. "Forbidden Roads: The Discriminatory West Bank Road Regime." (Jerusalem: B'Tselem). http://www.btselem.org/publications/summaries/200408_forbidden_roads.

Balfour, Arthur James. 1917. "The Balfour Declaration." *Israel Ministry of Foreign Affairs*. https://mfa.gov.il/mfa/foreignpolicy/peace/guide/pages/the%20balfour%20declaration.aspx.

Bar-Gal, Yoram. 1992. "Political Symbolization of an Urban Space: Naming Streets in Israel." *Horizons in Geography* 33: 119–32.

Bar-Gal, Yoram. 1993. *Homeland and Geography in a Hundred Years of Zionist Education*. Tel Aviv: Am Oved.

Bar-Gal, Yoram. 1999. *An Agent of Zionist Propaganda, The Jewish National Fund 1924–1947*. Jerusalem: Keter.

Bar-Gal, Yoram. 2002. "Maps and Nationalism: A New Reading of the Atlas of Israel." *Horizons in Geography* 55: 7–25.

Bar-Gal, Yoram. 2003a. *Propaganda and Zionist Education: The Jewish National Fund 1924–1947*. Rochester: University of Rochester Press.

Bar-Gal, Yoram. 2003b. "The Blue Box and JNF Propaganda Maps, 1930–1947." *Israel Studies* 8 (1): 1–19.

Bar-Gal, Yoram. 2004. "From Berlin to Jerusalem—Professor David Amiran and the Atlas of Israel." *Erdkunde* 58 (1): 31–41.

Bar-Gal, Yoram, and Bruria Bar-Gal. 2008. "'To Tie the Cords between the People and Its Land': Geography Education in Israel." *Israel Studies* 13 (1): 44–67.

Bar-Tal, Daniel, and Dikia Antebi. 1992. "Siege Mentality in Israel." *Productions Vives Sur Les Représentations Sociales* 1 (1): 49–67.

Barron, Rod. 2015. "Europe in 1915—German Perspectives on the War through Two Sets of Satirical Propaganda Maps." *Barron Maps*. http://www.barronmaps.com/europe-1915-satirical-propaganda-maps/.

Barthes, Roland. 1994. "Myth Today." In *Ideology*, edited by Terry Eagleton, 162–72. London: Longman.

Bazerman, Charles, and James G Paradis, eds. 1991. *Textual Dynamics of the Professions: Historical and Contemporary Studies of Writing in Professional Communities*. Madison: University of Wisconsin Press.

Ben-Eliezer, Uri. 1998. *The Making of Israeli Militarism*. Bloomington: Indiana University Press.

Ben-Israel, Arnon. 2003. "The Settlements' Cultural Landscape." Master's Thesis, Tel Aviv University.

Ben-Rafael Galanti, S., W. E. Aaronson, and Izhak Schnell. 2001. "Power and Changes in the Balance between Ideology and Pragmatism in the Right Wing Likud Party." *GeoJournal* 53 (3): 263–72.

Ben-Yehuda, Nachman. 1995. *The Masada Myth: Collective Memory and Mythmaking in Israel*. Madison: University of Wisconsin Press.

Ben-Yehuda, Nachman. 2009. "The History, Myth, and Science of Masada: The Making of an Historical Ethnography." In *Ethnographies Revisited: Constructing Theory in the Field*, edited by Antony J Puddephatt, William Shaffir, and Steven W Kleinknecht, 331–45. London: Routledge.

Ben-Ze'ev, Efrat. 2011. *Remembering Palestine in 1948: Beyond National Narratives*. Cambridge: Cambridge University Press.

Ben-Ze'ev, Efrat. 2012. "Mental Maps and Spatial Perceptions: The Fragmentation of Israel-Palestine". In *Mapping Cultures: Place, Practice, Performance*, edited by Les Roberts, 237–59. London: Palgrave Macmillan.

Benvenisti, Meron. 1989. *The Shepherds' War: Collected Essays (1981–1989)*. Jerusalem: West Bank Data Base Project.

Berger, Tamar. 2008. *In the Space between World and Playing: The Model in Israeli Culture*. Tel Aviv: Resling.

Bernstein, D. 1987. *Women in Eretz Israel: The Aspiration for Equality*. Tel Aviv: Hakibutz Hameuchad.

Bier, Jess. 2017. *Mapping Israel, Mapping Palestine: How Occupied Landscapes Shape Scientific Knowledge*. Cambridge, MA: MIT Press.

Biger, Gideon. 2002. "Village Boundaries as a Factor in Delimiting International Boundaries in the Middle East." In *The Razor's Edge: International Boundaries and Political Geography: Essays in Honour of Professor Gerald Blake*, edited by Clive H Schofield and Gerald Henry Blake, 463–71. London: Kluwer Law International.

Biger, Gideon. 2004. *The Boundaries of Modern Palestine: 1840–1947*. London: Routledge.

Biger, Gideon. 2008. "The Boundaries of Israel—Palestine Past, Present, and Future: A Critical Geographical View." *Israel Studies* 13 (1): 68–93.

Biger, Gideon. 2009. "Sovereignty in No Man's Land: Lessons from the 1949 Armistice Lines." *Presentation at International Boundaries Research Unit*. Durham University.

Billig, Michael. 1989. *Arguing and Thinking: A Rhetorical Approach to Social Psychology*. Cambridge: Cambridge University Press.

Bitan, H. 1992. "Government Naming Committee." *Land of Israel* 23: 366–71.

Bitan, H. 1998. *The Atlas of Settlements and Places in Israel*. Jerusalem: Carta.

Bittner, Christian, Georg Glasze, Boris Michel, and Cate Turk. 2011. "Krisen- Und Konflikt-Karten Im Web 2.0." *Geographische Rundschau* 63 (11): 60–65.

Bittner, Christian, Georg Glasze, and Cate Turk. 2013. "Tracing Contingencies: Analyzing the Political in Assemblages of Web 2.0 Cartographies." *GeoJournal* 78 (6): 935–48.

Black, Jeremy. 1997. *Maps and Politics*. Chicago: University of Chicago Press.

Bowker, Geoffrey, and Susan Leigh Star. 1999. *Sorting Things Out: Classification and Its Consequences*. Edited by Wiebe E. Bijker, W. B. Carlson, and T. Pinch. Cambridge, MA: MIT Press.

Braverman, Irus. 2014. *Planted Flags: Trees, Land and Law in Israel/Palestine.* Cambridge: Cambridge University Press.

Brawer, Moshe. 1978. "The Impact of Boundaries on Patterns of Rural Settlement: The Case of Samaria (Israel)." *GeoJournal* 2 (6): 539–47.

Brawer, Moshe. 1983. "Die Neubelebung Einer Internationalen Grenze Auf Der Sinai-Halbinsel." *Regio Basiliensis* 24 (1): 27–37.

Brawer, Moshe. 2002. "The Making of an Israeli-Palestinian Boundary." In *The Razor's Edge: International Boundaries and Political Geography: Essays in Honour of Professor Gerald Blake*, edited by Clive H Schofield and Gerald Henry Blake, 473–91. London: Kluwer Law International.

Brown, Barry, and Eric Laurier. 2005. "Maps and Journeys: An Ethnomethodological Investigation." *Cartographica: The International Journal for Geographic Information and Geovisualization* 40 (3): 17–33.

Buttimer, Anne, and Gerard Fahy. 1999. "Imagining Ireland through Geography Texts." In *Text and Image: Social Construction of Regional Knowledges*, edited by Anne Buttimer, Stanley D. Brunn, and Ute Wardenga, 179–91. Leipzig: Institut für Länderkunde.

Caplan, Jane, and John Torpey, eds. 2001. *Documenting Individual Identity: The Development of State Practices in the Modern World.* Princeton, NJ: Princeton University Press.

Carroll-Burke, Patrick. 2002. "Material Designs: Engineering Cultures and Engineering States—Ireland 1650–1900." *Theory and Society* 31 (1): 75–114.

Carroll, Patrick. 2006. *Science, Culture, and Modern State Formation.* Berkeley: University of California Press.

Carta. 1980. *The Carta Atlas for the History of the State of Israel.* Jerusalem: Carta.

Carta. 2009. *Israel (and Autonomous Areas) Physical Map 1:270,000.* Jerusalem: Carta.

Chappell, Bill. 2014. "Google Maps Display Crimean Border Differently in Russia, U.S." NPR. https://www.npr.org/sections/thetwo-way/2014/04/12/302337754/google-maps- displays-crimean-border-differently-in-russia-u-s.

Cohen, Saul B., and Nurit Kliot. 1981. "Israel's Place-Names as Reflection of Continuity and Change in Nation-Building." *Names* 29 (3): 227–48.

Cohen, Saul B., and Nurit Kliot. 1992. "Place-Names in Israel's Ideological Struggle over the Administered Territories." *Annals of the Association of American Geographers* 82 (4): 653–80.

Collier, Peter. 2006. "The Work of the British Government's Air Survey Committee and Its Impact on Mapping in the Second World War." *Photogrammetric Record* 21 (114): 100–109.

Collins-Kreiner, Noga, and Yoel Mansfeld. 2005. "Mapping of the Holy Land: Contemporary Religious Mapping." *Tijdschrift Voor Economische En Sociale Geografie* 96 (1): 105–20.

Collins-Kreiner, Noga, Y. Mansfeld, and N. Kliot. 2006. "The Reflection of a Political Conflict in Mapping: The Case of Israel's Borders and Frontiers." *Middle Eastern Studies* 42 (3): 381–408.

Collins, Harry M, ed. 1981. "Special Issue: Knowledge and Controversy: Studies of Modern Natural Science." *Social Studies of Sciences* 11 (1).

Collins, Harry M., and Trevor J. Pinch. 1993. *The Golem: What Everyone Should Know about Science.* Cambridge: Cambridge University Press.

Crampton, Jeremy W. 2001. "Maps as Social Constructions: Power, Communication and Visualization." *Progress in Human Geography* 25 (2): 235–52.

Crampton, Jeremy W. 2009. "Cartography: maps 2.0." *Progress in Human Geography* 33 (1): 91–100.
Crampton, Jeremy W. 2010a. *Mapping: A Critical Introduction to Cartography and GIS*. Chichester: Wiley-Blackwell.
Crampton, Jeremy W. 2010b. "Cartographic Calculations of Territory." *Progress in Human Geography* 35 (1): 92–103.
Crampton, Jeremy W., and John Krygier. 2006. "An Introduction to Critical Cartography." *ACME: An International Journal for Critical Geographies* 4 (1): 11–33.
Culcasi, Karen. 2006. "Cartographically Constructing Kurdistan within Geopolitical and Orientalist Discourses." *Political Geography* 25 (6): 680–706.
Davis, Diana. 2016. *The Arid Lands: History, Power, Knowledge*. Cambridge, MA: MIT Press.
Davis, Diana, and Edmund Burke III. 2011. *Environmental Imaginaries of the Middle East and North Africa*. Athens: Ohio University Press.
Davis, Rochelle. 2007. "Mapping the Past, Re-Creating the Homeland: Memories of Village Places in Pre-1948 Palestine." In *Nakba: Palestine, 1948, and the Claims of Memory*, edited by Ahmad H. Sa'di and Lila Abu-Lughod, 53–76. New York: Columbia University Press.
Day, David. 2008. *Conquest: How Societies Overwhelm Others*. Oxford: Oxford University Press.
DEA. 2012. "Interdisciplinary Research Is Key to Solving Society's Problems." Copenhagen.
Derzia, O. 1993. "Space and Names, Names of Settlements, National Processes and Social Attitudes." PhD Dissertation, Geography Department, Tel Aviv University.
Doumani, Beshara B. 1999. "Rediscovering Ottoman Palestine: Writing Palestinians into History." In *The Israel/Palestine Question: A Reader*, edited by Ilan Pappé, 10–35. New York: Routledge.
Dror, Yuval. 2004. "Textbook Images as a Means of 'Nation/State Building': Zionist Geographical Textbooks 1918 to 1948." *History of Education Review* 33 (2): 59–72.
Duani, Maya. 2010. "From Drainage to Conservation of the Huleh Wetlands: Tracing the Dynamics of Nature Intervention." PhD Dissertation, University of Haifa.
Durkheim, Émile. 1965. *The Elementary Forms of the Religious Life*. Translated by Joseph Ward Swain. New York: Free Press.
Edney, Matthew H. 1997. *Mapping an Empire: The Geographical Construction of British India, 1765–1843*. Chicago: University of Chicago Press.
Edwards, Tom. 2009. "Virtual Sovereignty, Online Cartography and the Battle for Public Mindshare." Presented at International Boundaries Research Unit, Durham University.
Eghigian, Greg, Andreas Killen, and Christine Leuenberger, eds. 2007. "The Self as Project: Politics and the Human Sciences in the Twentieth Century." *Osiris* 22.
El-Atrash, Ahmad. 2016. "Politics of Informal Urbanization and the Battle for Urban Rights in Jerusalem." *Jerusalem Quarterly* 65: 104–10.
El-Atrash, Ahmad, Hilmi S. Salem, and Jad E. Isaac. 2009. "Planning Towards a Sustainable Land Transportation System in a Future Palestinian State." *Journal of Nature Science and Sustainable Technology* 2 (3): 305–24.
El-Haj, Nadia Abu. 2002. "Producing (Arti) Facts: Archaeology and Power during the British Mandate of Palestine." *Israel Studies* 7 (2): 33–61.

Falah, Ghazi. 1989. "Israeli State Policy toward Bedouin Sedentarization in the Negev." *Journal of Palestine Studies* 18 (2): 71–91.

Falah, Ghazi. 1996. "The 1948 Israeli-Palestinian War and Its Aftermath: The Transformation and De-Signification of Palestine's Cultural Landscape." *Annals of the Association of American Geographers* 86 (2): 256–85.

Falah, Ghazi. 2002. "The 'Green Line' Revisited: October 2000." In *The Razor's Edge: International Boundaries and Political Geography: Essays in Honour of Professor Gerald Blake*, edited by Clive H. Schofield and Gerald Henry Blake, 493–511. London: Kluwer Law International.

Feige, Michael. 1999. "Yesha Is Here Occupied Territories Is Here: Scientific Practices and the Formation of Israeli Space." *Teoria Ve-Bikoret [Theory and Criticism]* 14: 111–31.

Ferguson, Niall. 2002. *Empire: How Britain Made the Modern World*. London: Penguin Books.

Fields, Gary. 2017. *Enclosure: Palestinian Landscapes in a Historical Mirror*. Oakland: University of California Press.

Finkelstein, Israel, Ze'ev Herzog, Lily Singer-Avitz, and David Ussishkin. 2007. "Has King David's Palace in Jerusalem Been Found?" *Tel Aviv: Journal of the Institute of Archaeology of Tel Aviv University* 34 (2): 142–64.

Fleishman, L., and A. Solomon. 2005. "To the Question Where Is the Green Line? The Answer: What Is the Green Line?" *Alpaim* 29: 26–52.

Forman, Geremy, and Alexandre (Sandy) Kedar. 2004. "From Arab Land to 'Israel Lands': The Legal Dispossession of the Palestinians Displaced by Israel in the Wake of 1948." *Environment and Planning D: Society and Space* 22 (6): 809–30.

Foster, Jeremy. 2003. "'Land of Contrasts' or 'Home We Have Always Known'? The SAR&H and the Imaginary Geography of White South African Nationhood, 1910–1930." *Journal of Southern African Studies* 29 (3): 657–80.

Frenkel, Michal, Yehouda Shenhav, and Hanna Herzog. 1997. "The Political Embeddedness of Managerial Ideologies in Pre-state Israel: The Case of PPL 1920–1948." *Journal of Management History (Archive)* 3 (2): 120–44.

Fried, Talia. 2012a. "Mapping the Nakba of 1948: The Case of the Israeli NGO Zochrot." Presentation at Association for Israel Studies, University of Haifa (unpublished).

Fried, Talia. 2012b. "Maps, Information Design, and Social Change: The Case of Destroyed Palestinian Villages." Paper Presented at the 4S Annual Meeting, Copenhagen Business School, Frederiksberg, Denmark (unpublished).

Frontline. 2002. "Shattered Dreams of Peace: The Road from Oslo." Public Broadcasting Service (PBS). https://www.pbs.org/wgbh/pages/frontline/shows/oslo/etc/script.html.

Galnoor, Itzhak. 1995. *The Partition of Palestine: Decision Crossroads in the Zionist Movement*. Albany: State University of New York Press.

Garb, Yaakov. 2004. "Constructing the Trans-Israel Highway's Inevitability." *Israel Studies* 9 (2): 180–217.

Gavish, Dov. 2005. *A Survey of Palestine under the British Mandate*. New York: Routledge.

Gavish, Dov, and Ruth Kark. 1993. "The Cadastral Mapping of Palestine, 1858–1928." *Geographical Journal* 159 (1): 70–80.

Geertz, Clifford. 1973. *The Interpretation of Cultures*. New York: Basic Books.

Geertz, Clifford. 1994. "Ideology as a Cultural System." In *Ideology*, edited by Terry Eagleton, 279–94. London: Longman.

Gellner, Ernest. 1983. *Nations and Nationalism*. Ithaca, NY: Cornell University Press.

Gerovitch, Slava. 2001. "'Mathematical Machines' of the Cold War: Soviet Computing, American Cybernetics and Ideological Disputes in the Early 1950s." *Social Studies of Science* 31 (2): 253–87.

Giddens, Anthony. 1984. *The Constitution of Society: Outline of the Theory of Structuration*. Berkeley: University of California Press.

Gilbert, G. Nigel, and Michael Mulkay. 1984. *Opening Pandora's Box: A Sociological Analysis of Scientists' Discourse*. New York: Cambridge University Press.

Goren, Haim. 2002. "Sacred, but Not Surveyed: Nineteenth-Century Surveys of Palestine." *Imago Mundi* 54: 87–110.

Gramsci, Antonio. 1971. *Selections from the Prison Notebooks of Antonio Gramsci*. Edited and translated by Quintin Hoare and Geoffrey Nowell-Smith. New York: International Publishers.

Greenberg, Raphael. 2009. "Towards an Inclusive Archaeology in Jerusalem: The Case of Silwan/The City of David." *Public Archaeology* 8 (1): 35–50.

Günthner, Susanne, and Hubert Knoblauch. 1995. "Culturally Patterned Speaking Practices—The Analysis of Communicative Genres." *Pragmatics* 5 (1): 1–32.

Gurevitz, Zali. 2007. *On the Place*. Tel Aviv: Am Oved.

Habiballah, Nahed. 2016. "The Unmaking of Arab Jerusalem through Settlement Construction." *Jerusalem Quarterly* 65: 111–17.

Hacking, Ian. 1990. *The Taming of Chance*. Cambridge: Cambridge University Press.

Hall, Stuart. 1986. "Gramsci's Relevance for the Study of Race and Ethnicity." *Journal of Communication Inquiry* 10 (2): 5–27.

Halper, Jeff. 2004. "Beyond Road Maps & Walls." *The Link* 37 (1): 1–13. https://www.kairoscanada.org/wp-content/uploads/2012/02/TheLink_BeyondWalls__1_2004.pdf.

Harley, John B. 1991. "Deconstructing the Map." In *Writing Worlds: Discourse, Text and Metaphor in the Representation of Landscape*, edited by Barnes Trevor and James S. Duncan, 231–47. London: Routledge.

Harvey, Francis. 2009. "Maps and Governance." In *International Encyclopedia of Human Geography*, edited by Rob Kitchin and Nigel Thrift, Volume 6. Oxford: Elsevier.

Hasson, S., ed. 2012. *The Production of Space in Israel: The Map of Settlement an Land*. Jerusalem: Keter (in Hebrew).

Heffernan, Michael. 1996. "Geography, Cartography and Military Intelligence: The Royal Geographical Society and the First World War." *Transactions of the Institute of British Geographers* 21 (3): 504–33.

Herb, Guntram Henrik. 1997. *Under the Map of Germany—Nationalism and Propaganda 1918–1945*. London: Routledge.

Herman, Ellen. 1995. *The Romance of American Psychology: Political Culture in the Age of Experts*. Berkeley: University of California Press.

Herzog, Hanna. 1999. *Gendering Politics: Women in Israel*. Michigan: University of Michigan Press.

Hobsbawn, Eric J. 1992. *Nations and Nationalism since 1780: Programme, Myth, Reality*. Cambridge: Cambridge University Press.

Hoffman, I. 2002. "Between National History and General History." In *Values and Aims in Israel's Study Curricula*, edited by I. Hoffman and Izhak Schnell. Tel Aviv: Achs.

Horowitz, Dan. 1989. "Before the State: Communal Politics in Palestine under the Mandate." In *The Israeli State and Society: Boundaries and Frontiers*, edited by Baruch Kimmerling, 28–65. Albany: State University of New York Press.

Horowitz, Dan, and Moshe Lissak. 1989. *Trouble in Utopia: The Overburdened Polity of Israel*. Albany: State University of New York Press.

Jackson, John Brinckerhoff. 1970. "The Public Landscape." In *Landscapes: Selected Writings of J. B. Jackson*, edited by Ervin H. Zube, 153–60. Amherst: University of Massachusetts Press.

Jacob, Christian. 2005. *The Sovereign Map: Theoretical Approaches in Cartography throughout History*. Edited by Edward H. Dahl. Translated by Tom Conley. Chicago: University of Chicago Press.

Jamal, Amal. 2006. "The Arab Leadership in Israel: Ascendance and Fragmentation." *Journal of Palestine Studies* 35 (2): 6–22.

Jasanoff, Sheila. 2004. *States of Knowledge: The Co-Production of Science and the Social Order*. New York: Routledge.

Jasanoff, Sheila, Gerald E. Markle, James C. Peterson, and Trevor J. Pinch, eds. 1995. *Handbook of Science and Technology Studies*. London: Sage.

Jones, Reece. 2008. "Categories, Borders and Boundaries." *Progress in Human Geography* 33 (2): 174–89.

Kadmon, Naftali. 2004. "Toponymy and Geopolitics: The Political Use—and Misuse—of Geographical Names." *Cartographic Journal* 41 (2): 85–87.

Kark, Ruth. 1995. *Land and Settlement in Eretz Israel 1830–1990: Selected Papers by Prof. Ruth Kark*. Jerusalem: Land-Use Research Institute.

Kaufman, Asher. 2002. "Who Owns the Shebaa Farms? Chronicle of a Territorial Dispute." *Middle East Journal* 56 (4): 576–95.

Kaufman, Asher. 2009. "'Let Sleeping Dogs Lie:' On Ghajar and Other Anomalies in the Syria-Lebanon-Israel Tri-Border Region." *Middle East Journal* 63 (4): 539–60.

Kedar, Alexandre (Sandy). 1996. "Israeli Law and the Redemption of Arab Land 1948–1969." PhD Dissertation, Harvard Law School.

Kedar, Alexandre (Sandy). 2001. "The Legal Transformation of Ethnic Geography: Israeli Law and the Palestinian Landholder 1948–1967." *New York University Journal of International Law and Politics*, 923: 923–1000.

Kellerman, Aharon. 1993. *Society and Settlement: Jewish Land of Israel in the Twentieth Century*. Albany: State University of New York Press.

Kemp, Adriana. 1999. "The Frontier Idiom on Borders and Territorial Politics in Post-1967 Israel." *Geography Research Forum* 19 (February): 78–97.

Kestler-D'Amours, Jillian. 2012. "The End of the Bedouin." *Le Monde Diplomatique* (Aug). https://mondediplo.com/2012/08/06bedouin.

Khalidi, Walid. 2006. *All That Remains: The Palestinian Villages Occupied and Depopulated by Israel in 1948*. Edited by Institute for Palestine Studies. Washington, DC.

Kimmerling, Baruch. 1983a. *Zionism and Territory: The Socio-Territorial Dimensions of Zionist Politics*. Berkeley: University of California Press.

Kimmerling, Baruch. 1983b. "Making Conflict a Routine: Cumulative Effects of the Arab-Jewish Conflict upon Israeli Society." *Journal of Strategic Studies* 6 (3): 13–45.

Kimmerling, Baruch. 1998. "Between Hegemony and Dormant Kulturkampf in Israel." *Israel Affairs* 4 (3–4): 49–72.

Kimmerling, Baruch. 2001. *The Invention and Decline of Israeliness: State, Society, and the Military*. Berkeley: University of California Press.

Kimmerling, Baruch, and Joel S Migdal. 2003. *The Palestinian People: A History*. Cambridge, MA: Harvard University Press.

Kitchin, Rob, and Martin Dodge. 2007. "Rethinking Maps." *Progress in Human Geography* 31 (3): 331–44.

Knorr Cetina, Karin. 1995. "Laboratory Studies: The Cultural Approach to the Study of Science." In *Handbook of Science and Technology Studies*, edited by Sheila Jasanoff, Gerald E. Markle, James C. Peterson, and Trevor J. Pinch, 140–66. London: Sage.

Konrad, Victor. 2011. "Borders, Bordered Lands and Borderlands: Geographical States of Insecurity between Canada and the United States and the Impacts of Security Primacy." Presentation at Fences, Walls and Borders State of Insecurity. Montréal: Université du Québec à Montréal.

Koopman, Sara. 2011. "Alter-Geopolitics: Other Securities Are Happening." *Geoforum* 42 (3): 274–84.

Kosonen, Katariina. 2008. "Making Maps and Mental Images: Finnish Press Cartography in Nation-Building, 1899–1942." *National Identities* 10 (1): 21–47.

Kretzmer, David. 2012. "The Law of Belligerent Occupation as a System of Control: Dressing Up Exploitation in Respectable Garb." In *The Impacts of Lasting Occupation: Lessons from Israeli Society*, edited by Daniel Bar-Ta and Izhak Schnell, 31–60. Oxford: Oxford University Press.

Krishna, Sankaran. 1996. "Cartographic Anxiety: Mapping the Body Politic in India." In *Challenging Boundaries: Global Flows, Territorial Identities*, edited by Michael J. Shapiro and Hayward R. Alker, 193–215. Minneapolis: University of Minnesota Press.

Lamm, Z. 2002. *In the Confusion of Ideologies*. Jerusalem: Magness.

Leibler, Anat E. 2004. "Statisticians' Ambition: Governmentality, Modernity and National Legibility." *Israel Studies* 9 (2): 121–49.

Lentin, Ronit. 2007. "The Memory of Dispossession, Dispossessing Memory: Israeli Networks Commemorating the Nakba." In *Performing Global Networks*, edited by Karen Fricker and Ronit Lentin, 206–27. Newcastle: Cambridge Scholars.

Leuenberger, Christine. 2006. "Constructions of the Berlin Wall: How Material Culture Is Used in Psychological Theory." *Social Problems* 53 (1): 18–37.

Leuenberger, Christine. 2011. "From the Berlin Wall to the West Bank Barrier: How Material Objects and Psychological Theories Can Be Used to Construct Individual and Cultural Traits." In *After the Berlin Wall: Germany and Beyond*, edited by Katharina Gerstenberger and Jana Evans Braziel, 59–83. New York: Palgrave Macmillan.

Leuenberger, Christine. 2012. "Mapping Israel/Palestine: Constructing National Territories across Different Online International Newspapers." *Bulletin du Centre de Recherche Français à Jérusalem* 23.

Leuenberger, Christine. 2013a. "The Rhetoric of Maps: International Law as a Discursive Tool in Visual Arguments." *Law and Ethics of Human Rights* 7 (1). https://doi.org/10.1515/lehr-2013-0002.

Leuenberger, Christine. 2013b. "Map-Making for Palestinian State-Making." *Arab World Geographer* 16 (1): 54–74.

Leuenberger, Christine. 2014. "Technologies, Practices and the Reproduction of Conflict: The Impact of the West Bank Barrier on Peace Building." In *Borders, Fences and Walls: State of Insecurity?*, edited by Elisabeth Vallet, 211–30. Burlington, VT: Ashgate.

Leuenberger, Christine. 2016a. "Mapping Divided Cities and Their Separation Walls: Berlin and Jerusalem." *Jerusalem Quarterly* 65: 86–103.

Leuenberger, Christine. 2016b. "Maps as Politics: Mapping the West Bank Barrier." *Journal of Borderlands Studies* 31 (3): 339–364.

Leuenberger, Christine, and Ahmad El-Atrash. 2015. "Building a Neoliberal Palestinian State under Closure: The Economic and Spatial Implications of Walls and Barriers." *Economic Sociology European Electronic Newsletter* 16 (2): 21–31.

Leuenberger, Christine, and Izhak Schnell. 2010. "The Politics of Maps: Constructing National Territories in Israel." *Social Studies of Science* 40 (6): 803–42.

Levy, Gal, and Orna Sasson-Levy. 2008. "Militarized Socialization, Military Service, and Class Reproduction: The Experiences of Israeli Soldiers." *Sociological Perspectives* 51 (2): 349–74.

Levy, Gideon. 2015. "Religious Ultranationalist Zionists Have Taken Over Israel." *Haaretz*. Jerusalem. https://www.haaretz.com/opinion/.premium-religious-zionists-have-won-israel-1.5381210.

Lewis, Peter G. 1982. "The Politics of Iranian Place-Names." *Geographical Review* 72 (1): 99–102.

Liebman, Charles S., and Eli'ezer Don-Yiḥya. 1983. *Civil Religion in Israel: Traditional Judaism and Political Culture in the Jewish State*. Berkeley: University of California Press.

Liphschitz, Nili, and Gideon Biger. 2004. *Green Dress for a Country: Afforestation in Eretz Israel: The First Hundred Years 1850–1950*. Jerusalem: KKL—Land Use Research Institute, Ariel Publishing House.

Lissak, Moshe. 1983. "Paradoxes of Israeli Civil-Military Relations: An Introduction." *Journal of Strategic Studies* 6 (3): 1–12.

Luckmann, Thomas. 1986. "Grundformen Der Gesellschaftlichen Vermittlung Des Wissens: Kommunikative Gattungen." In *Kultur Und Gesellschaft: Kölner Zeitschrift Für Soziologie Und Sozialpsychologie*, edited by Friedhelm Neidhardt and M. Rainer Lepsius, 191–211. Opladen: Westdeutscher Verlag.

Manoff, Einat. 2011. "Counter-Mapping Return." *Sedek: A Journal on the Ongoing Nakba* 6: 1–11.

Mautner, Menachem. 2008. *Law and Culture in Israel at the Threshold of the Twenty-First Century*. Tel Aviv: Am Oved.

McCloskey, Deirdre N. 1990. *If You're so Smart: The Narrative of Economic Expertise*. Chicago: University of Chicago Press.

McGirk, Tim. 2010. "Archaeology in Jerusalem: Digging Up Trouble." *Time Magazine*, February. http://content.time.com/time/magazine/article/0,9171,1957350,00.html.

McNally, Mark, and John Schwarzmantel, eds. 2009. *Gramsci and Global Politics: Hegemony and Resistance*. London: Routledge.

Medzini, A. 2011. "The Use of Maps and Atlases in Constituting National Awareness: Israel vs. the Palestinian Authority." In *The Political Role of Maps in the Israeli-Palestinian Conflict*, edited by Izhak Schnell and Ephraim Lavie, 57–78. Tel Aviv: Steinmetz Center for Peace, Tel Aviv University.

Megoran, Nick. 2011. "War and Peace? An Agenda for Peace Research and Practice in Geography." *Political Geography* 30 (4): 178–89.

Merry, Sally Engle. 2006. "Transnational Human Rights and Local Activism: Mapping the Middle." *American Anthropologist* 108 (1): 38–51. http://www.jstor.org/stable/3804730.

Mezzadra, Sandro, and Brett Neilson. 2013. *Border as* Method. Durham: Duke University Press.

Michael, Shalev. 1989. "Jewish Organized Labor and the Palestinians: A Study of State/Society Relations in Israel." In *The Israeli State and Society: Boundaries and Frontiers*, edited by Baruch Kimmerling, 93–133. Albany: State University of New York Press.

Migdal, Joel S. 1989. "The Crystallization of the State and the Struggles over Rulemaking: Israel in Comparative Perspective." In *The Israeli State and Society: Boundaries and Frontiers*, edited by Baruch Kimmerling, 1–27. Albany: State University of New York Press.

Miles, William F. S. 2011. "Border Pedagogy in Israel." *Middle East Journal* 65 (2): 253–77.

Mizrachi, Yonathan. 2011. "Archeology Should Not Be Used as a Weapon in the Struggle over Jerusalem." *Palestine-Israel Journal* 17 (1): 96–98.

Monmonier, Mark. 1996. *How to Lie with Maps*. Chicago: University of Chicago Press.

Monmonier, Mark. 2006. "Cartography: Uncertainty, Interventions, and Dynamic Display." *Progress in Human Geography* 30 (3): 373–81.

MOPIC. 1998. "Regional Plan for the West Bank Governorates." Ramallah, West Bank.

Morris, Benny. 1989. *Birth of the Palestinian Refugee Problem*. Cambridge: Cambridge University Press.

Morton, Ella. 2015. "U-Boats and Octopuses Collide in These WWI Propaganda Maps." *Atlas Obscura*. https://www.atlasobscura.com/articles/uboats-and-octopuses-collide-in-these-wwi-propaganda-maps.

Mukerji, Chandra. 2002. "Material Practices of Domination: Christian Humanism, the Built Environment, and Techniques of Western Power." *Theory and Society* 31 (1): 1–34.

Naor, A. 2001. *The Greater Eretz Israel*. Haifa: Haifa University Press.

Nelkin, Dorothy. 1995. "Science Controversies: The Dynamics of Public Disputes in the United States." In *Handbook of Science and Technology Studies*, edited by Sheila Jasanoff, Gerald E. Markle, James C. Peterson, and Trevor J. Pinch, 444–56. London: Sage.

Nelson, John S., Allan Megill, and Donald N. McCloskey. 1987. "Rhetoric of Inquiry." In *Rhetoric of the Human Sciences: Language and Argument in Scholarship and Public Affairs*, edited by John S. Nelson, Allan Megill, and Donald N. McCloskey, 3–18. Madison: University of Wisconsin Press.

Newman, David. 2001. "From National to Post-National Territorial Identities in Israel-Palestine." *GeoJournal* 53 (3): 235–46.

Newman, David. 2005. "From Hitnachalut to Hitnatkut: The Impact of Gush Emunim and the Settlement Movement on Israeli Politics and Society." *Israel Studies* 10 (3): 192–224.

Newman, David. 2008. "The Formation of National Identity in Israel/Palestine: The Construction of Spatial Knowledge and Contested Territorial Narratives." In *Promoting Conflict or Peace through Identity*, edited by Nikki Slocum-Bradley, 61–79. Hampshire: Ashgate.

Newman, David, and Anssi Paasi. 1998. "Fences and Neighbours in the Postmodern World: Boundary Narratives in Political Geography." *Progress in Human Geography* 22 (2): 186–207.

Nolte, A. and H. Yacobi. 2015. "Politics, Infrastructure and Representation: The Case of Jerusalem's Light Rail." *Cities* 43: 28–36.

OECD. 2009. "Concepts and Dilemmas of State Building in Fragile Situations: From Fragility to Resilience." *OECD Journal on Development* 9 (3): 61–148.

Paasi, Anssi. 1999. "The Changing Pedagogies of Space: Representation of the Other in Finnish School Geography Textbooks." In *Text and Image: Social Construction of Regional Knowledges*, edited by Anne Buttimer, Stanley D. Brunn, and Ute Wardenga, 226–37. Leipzig: Institut für Länderkunde.

Paasi, Anssi. 2002. "Regional Transformation in the European Context: Notes on Regions, Boundaries and Identity." *Space and Polity* 6 (2): 197–201.

Paasi, Anssi, and Prokkola, E-K. 2008. "Territorial Dynamics, Cross-Border Work and Everyday Life in the Finnish-Swedish Border Area." *Space and Polity* 12 (1): 13–29.

Palestinian Authority (PA). 2008. *The NGO Mapping Project: A New Approach to Advancing Palestinian Civil Society*. Ramallah, West Bank: Center for the Study of the Presidency.

Pappe, Ilan. 1995. "Critique and Agenda: The Post-Zionist Scholars in Israel." *History and Memory* 7 (1): 66–90.

Peri, Yoram. 1983. "Party-Military Relations in a Pluralist System." *Journal of Strategic Studies* 6 (3): 46–63.

Perkins, Chris. 2004. "Cartography—Cultures of Mapping: Power in Practice." *Progress in Human Geography* 28 (3): 381–91.

Pickles, John. 1992. "Texts, Hermeneutics and Propaganda Maps." In *Writing Worlds: Discourse, Text, and Metaphor in the Representation of Landscape*, edited by Trevor J. Barnes and James S. Duncan, 193–230. London: Routledge.

Pickles, John. 2004. *A History of Spaces: Cartographic Reason, Mapping and the Geo-Coded World*. New York: Routledge.

Pinch, Trevor J. 1985. "Towards an Analysis of Scientific Observation: The Externality and Evidential Significance of Observational Reports in Physics." *Social Studies of Science* 15 (1): 3–36.

Pinch, Trevor J. 1996. "Rhetoric and the Cold Fusion Controversy: From the Chemists' Woodstock to the Physicists' Altamont." In *Science, Reason and Rhetoric*, edited by Trevor Melia, J. E. McGuire, and Henry Krips, 153–76. Pittsburgh: University of Pittsburgh Press.

Podeh, Elie. 2002. *The Arab-Israeli Conflict in Israeli History Textbooks, 1948–2000*. London: Bergin & Garvey.

Popescu, Gabriel. 2012. *Bordering and Ordering the Twenty-First Century: Understanding Borders*. New York: Rowman & Littlefield.

Porter, Libby, and Oren Yiftachel. 2019. "Urbanizing Settler-Colonial Studies: Introduction to the Special Issue." *Settler Colonial Studies* 9 (2): 177–86.

Portugali, Juval. 1991. "Jewish Settlement in the Occupied Territories: Israel's Settlement Structure and the Palestinians." *Political Geography Quarterly* 10 (1): 26–53.

Portugali, Juval. 1993. *Implicate Relations: Society and Space in the Israeli-Palestinian Conflict*. Dordrecht: Kluwer Acadamic.

Radcliffe, Sarah A. 2009. "National Maps, Digitalisation and Neoliberal Cartographies: Transforming Nation-State Practices and Symbols in Postcolonial Ecuador." *Transactions of the Institute of British Geographers* 34 (4): 426–44.

Radcliffe, Sarah A. 2010. "Re-Mapping the Nation: Cartography, Geographical Knowledge and Ecuadorean Multiculturalism." *Journal of Latin American Studies* 42 (2): 293–323.

Raj, Kapil. 2007. *Relocating Modern Science: Circulation and the Construction of Knowledge in South Asia and Europe, 1650–1900*. New York: Palgrave Macmillan.

Ram, Uri. 1999. "The Colonization Perspective in Israeli Sociology." In *The Israel/Palestine Question*, edited by Ilan Pappé, 55–80. New York: Routledge.

Rebhun, Uzi, and Chaim Isaac Waxman, eds. 2004. *Jews in Israel: Contemporary Social and Cultural Patterns*. London: University Press of New England.

Rettig, Haviv, and Sheera Clare Frenkel. 2006. "Green Line in School Maps Angers Right-Wingers." *Jerusalem Post*. Jerusalem. https://www.jpost.com/Israel/Green-Line-in-school-maps-angers-right-wingers-43684.

Roberts, Jo. 2013. *Contested Land, Contested Memory: Israel's Jews and Arabs and the Ghosts of Catastrophe*. Toronto: Dundurn.

Robinson, Arthur H. 1952. *The Look of Maps: An Examination of Cartographic Design*. Madison: University of Wisconsin Press.

Rose-Redwood, Reuben, Derek Alderman, and Maoz Azaryahu. 2010. "Geographies of Toponymic Inscription: New Directions in Critical Place-Name Studies." *Progress in Human Geography* 34 (4): 453–70.

Routledge, Paul. 1996. "Critical Geopolitics and Terrains of Resistance." *Political Geography* 15 (6): 509–31.

Rudwick, Martin J. S. 1976. "The Emergence of a Visual Language for Geological Science 1760–1840." *History of Science* 14: 149–95.

Saarinen, Tomas Frederick. 1988. "Centering of Mental Maps of the World." *National Geographic Research* 4 (1): 112–27.

Salameh, Estephan. 2008. "The Implications for Public Land Management on Building and Developing the Palestinian State," *Ministry of Planning*. https://scholar.najah.edu/sites/default/files/conference-paper/implications-public-land-management-building-and-developing-palestinian-state.pdf.

Scanteam, and ARIJ. 2009. *Draft Report: End Review of the Physical Planning and Institution Building (PPIB) Project, Palestinian Territory*. Bethlehem: Applied Research Institute—Jerusalem (ARIJ).

Schnell, Izhak. 1999. "Narratives and Styles in Regional Geography of Israel." In *Text and Image: Social Construction of Regional Knowledges*, edited by Anne Buttimer, Stanley D. Brunn, and Ute Wardenga. Leipzig: Institut für Länderkunde.

Schnell, Izhak. 2000. "Heimatkunde in Israeli Regional Geography." In *Identity and Education in Germany and Israel*, edited by A. Hoffman, Izhak Schnell, G. Stephan, G. Brauer, and P. Fenn. Hamburg: Verlag Dr. Kovac.

Schnell, Izhak. 2001. "Transformation in Territorial Concepts: From Nation Building to Concessions." *GeoJournal* 53 (3): 221–34.

Schnell, Izhak. 2002. "Geography as an Educational Challenge." In *Values and Aims in Israel's Study Curricula*, edited by I. Hoffman and Izhak Schnell. Tel Aviv: Achs.

Schnell, Izhak. 2004. "Israeli Geographers in Search of a National Identity." *Professional Geographer* 56 (4): 560–73.

Schnell, Izhak. 2012. "Geographical Ramifications of the Occupation on Israeli Society." In *The Impacts of Lasting Occupation: Lessons from Israeli Society*, edited by Daniel Bar-Ta and Izhak Schnell, 93–121. Oxford: Oxford University Press.

Schnell, Izhak, and Ephraim Lavie, eds. 2012. *Map Discourse in the Israeli-Palestinian Conflict*. Tel Aviv: Steinmetz Center for Peace, Tel Aviv University.

Schnell, Izhak, and Christine Leuenberger. 2014. "Mapping Genres and Geopolitics: The Case of Israel." *Transactions of the Institute of British Geographers* 39 (4): 518–31.

Schnell, Izhak, and Shaul Mishal. 2008. "Place as a Source of Identity in Colonizing Societies: Israeli Settlements in Gaza." *Geographical Review* 98 (2): 242–59.

Schnell, Izhak, and Anda Rosenberg. 2011. "'We Shall Dress You in a Robe of Cement and Concrete:' How Discourse Concerning Megaengineering Projects Has Been Changing in Israel BT–Engineering Earth: The Impacts of Megaengineering Projects." In *Engineering Earth: The Impacts of Megaengineering Projects*, edited by Stanley D. Brunn, 663–81. Dordrecht: Springer.

Schwarzmantel, John. 2009. “Introduction: Gramsci in His Time and in Ours.” In *Gramsci and Global Politics: Hegemony and Resistance*, edited by Mark McNally and John Schwarzmantel, 1–16. London: Routledge.

Schweid, Eliezer. 1995. *The Idea of Judaism and a Culture*. Tel Aviv: Am Oved.

Scott, James C. 1987. “Resistance without Protest and without Organization: Peasant Opposition to the Islamic Zakat and the Christian Tithe.” *Comparative Studies in Society and History* 29 (3): 417–52.

Scott, James C. 1998. *Seeing like a State: How Certain Schemes to Improve the Human Condition Have Failed*. New Haven, CT: Yale University Press.

Scott, James C., John Tehranian, and Jeremy Mathias. 2002. “The Production of Legal Identities Proper to States: The Case of the Permanent Family Surname.” *Comparative Studies in Society and History* 44 (1): 4–44.

Scott, Pam, Evelleen Richards, and Brian Martin. 1990. “Captives of Controversy: The Myth of the Neutral Social Researcher in Contemporary Scientific Controversies.” *Science, Technology, & Human Values* 15 (4): 474–94.

Senge, Peter, and Otto Scharmer. 2001. “Community Action Research: Kearning as a Community of Practitioners, Consultants and Researchers.” In *Handbook of Action Research: Participative Inquiry and Practice*, edited by Peter Reason and Hilary Bradbury, 238–49. London: Sage.

Shafir, Gershon. 1999. “Zionism and Colonialism: A Comparative Approach.” In *The Israel/Palestine Question: A Reader*, edited by Ilan Pappé, 72–85. New York: Routledge.

Shalev, Aryeh. 1989. *Cooperation under the Shadow of Conflict: The Israeli-Syrian Armistice Regime, 1949–1955*. Tel Aviv: Ma'arachot.

Shalev, Chemi. 2014 “Berlin, 1933 and Jerusalem, 2014: When Racist Thugs Are on the Prowl,” July 2. Available at http:// www.haaretz.com/ blogs/ west-of-eden/.premium-1.602697.

Shatz, Adam, and Roane Carey. 2004. “Pastrami & Champagne.” *The Nation*, April. https://www.thenation.com/article/archive/pastrami-champagne/.

Shehadeh, Raja. 1997. “Land and Occupation: A Legal Review.” *Palestine-Israel Journal* 4 (2). https://pij.org/articles/473/land-and-occupation-a-legal-review.

Sheleg, Yair. 2000. *The New Religious*. Tel Aviv: Keter.

Smith, Anthony D. 1991. *National Identity*. Las Vegas: University of Nevada Press.

Soffer, Arnon. 1989. “Demography and the Shaping of Israel's Borders.” *Contemporary Jewry* 10 (2): 91–105.

Solel, Asher. 1988. “ישראל בארץ קרקעות על הזכויות גבולות של הגיאוגרפי התיחום ומיפויים.” ידוע לא לאור מוציא.

Solel, Asher. 1991. *The Geographical Delineation of the Boundaries and Rights for Lands in the Land of Israel*. Tel Aviv: Israeli Cartographic Center.

Sparke, Matthew. 1998. “A Map That Roared and an Original Atlas: Canada, Cartography, and the Narration of Nation.” *Annals of the Association of American Geographers* 88 (3): 463–95.

Srebro, Haim. 2009. “A Status Report of the Activity of the Survey of Israel.” *International Federation of Surveyors* (March). https://www.fig.net/resources/monthly_articles/2009/ march_2009/srebro_march_2009.pdf.

Srebro, Haim. 2011. "Cartographic Issues in Maps with Political Meanings." In *The Political Role of Maps in the Israeli-Palestinian Conflict*, edited by Izhak Schnell and Ephraim Lavie, 21–42. Tel Aviv: Steinmetz Center for Peace, Tel Aviv University.

Srebro, Haim, Ron Adler, and Dov Gavish. 2009. *60 Years of Surveying and Mapping Israel, 1948–2008*. Tel Aviv: Survey of Israel.

Srebro, Haim, and Maxim Shoshani. 2007. "The Order of Precedence of Boundary Delimitations." In *Strategic Integration of Surveying Services, FIG Working Week 2007*. Hong Kong SAR, China: International Federation of Surveyors (FIG). https://www.fig.net/resources/proceedings/fig_proceedings/fig2007/papers/ts_1g/ts01g_01_srebro_shoshani_1295.pdf.

Sterling, Brent L. 2009. *Do Good Fences Make Good Neighbors? What History Teaches Us about Strategic Barriers and International Security*. Washington, DC: Georgetown University Press.

Sternhell, Zeev. 1999. *The Founding Myths of Israel: Nationalism, Socialism, and the Making of the Jewish State*. Translated by David Maisel. Princeton, NJ: Princeton University Press.

Storey, John. 2009. *Cultural Theory and Popular Culture: An Introduction*. London: Longman.

Survey of Israel. 2009. *The New Atlas of Israel: The National Atlas*. Jerusalem: Survey of Israel.

Sutherland, Claire. 2005. "Nation-Building through Discourse Theory." *Nations and Nationalism* 11 (2): 185–202.

Szancer, Carmela. 2001. *The Official Mapping of the Land of Israel in the Transition from the British Mandate to the State of Israel during the Years 1945–1955*. Tel Aviv: Department of Geography and Human Environment, Tel Aviv University.

Tarrow, Sidney. 2013. *The Language of Contention: Revolutions in Words, 1688–2012*. Cambridge: Cambridge University Press.

Turnbull, David, and Helen Watson. 1989. *Maps Are Territories: Science Is an Atlas: A Portfolio of Exhibits*. Chicago: University of Chicago Press.

Tyner, Judith Ann. 1974. "Persuasive Cartography: An Examination of the Map as a Subjective Tool of Communication." PhD Dissertation, University of California, Los Angeles.

UN-Habitat. 2015. *Right to Develop: Planning Palestinian Communities in East Jerusalem*. East Jerusalem: United Nations Human Settlements Programme (UN-Habitat). https://www.un.org/unispal/document/auto-insert-196274/.

UN-OCHA. 2007. "The Humanitarian Impact on Palestinians of Israeli Settlements and Other Infrastructure in the West Bank." Jerusalem. https://www.ochaopt.org/content/humanitarian-impact-palestinians-israeli-settlements-and-other-infrastructure-west-bank.

UN Press Release. 2014. "Implementing Palestinian National Development Plan under Occupation 'an Uphill Struggle,' United Nations Seminar Told." *United Nations: Meetings Coverage and Press Releases*. https://www.un.org/press/en/2014/gapal1306.doc.htm.

UNRWA. 2012. "Barrier Impacts on Rural Communities: Rural Refugee Livelihoods under Pressure." Ramallah, West Bank.

Vallet, Elisabeth. 2014. *Borders, Fences and Walls: State of Insecurity?* Burlington, VT: Ashgate.

Verdery, Katherine. 1994. "The Elasticity of Land: Problems of Property Restitution in Transylvania." *Slavic Review* 53 (4): 1071–109.

Vertesi, Janet. 2008. "Mind the Gap: The London Underground Map and Users' Representations of Urban Space." *Social Studies of Science* 38 (1): 7–33.

Weiss, Efrat. 2006. "Settlers Reject Tamir's Textbooks." *Ynetnews*. Tel Aviv. https://www.ynetnews.com/articles/0,7340,L-3336338,00.html.

Weiss, Efrat. 2009. "Settlers Hand Out Fliers: Jews in West Bank—No Rockets in Tel Aviv." *Ynetnews*. Tel Aviv. https://www.ynetnews.com/articles/0,7340,L-3658648,00.html.

Weiss, Elliott. 2010. "Establishing Roots at Israel's Ben Gurion Airport Garden: Landscapes of National Identity." *National Identities* 12 (2): 199–210.

Weizman, Eyal. 2007. *Hollow Land: Israel's Architecture of Occupation*. New York: Verso.

Winichakul, Thongchai. 1994. *Siam Mapped: A History of the Geo-Body of a Nation*. Honolulu: University of Hawaii Press.

Wood, Denis, and John Fels. 1992. *The Power of Maps*. New York: Guilford Press.

Wood, Denis, and John Fels. 2008. "The Natures of Maps: Cartographic Constructions of the Natural World." *Cartographica* 43 (3): 189–202.

Wood, Denis, John Fels, and John Krygier. 2010. *Rethinking the Power of Maps*. New York: Guilford Press.

Wood, Denis, and John Krygier. 2005. *Making Maps: A Visual Guide to Map Design for GIS*. New York: Guilford Press.

World Bank. 2009. "West Bank and Gaza—Land Administration Project." Washington, DC. http://documents.worldbank.org/curated/en/369641468135025898/West-Bank-and-Gaza-Land-Administration-Project.

World Bank. 2016. "West Bank and Gaza—Seventh Palestinian National Development Plan Policy Grant Program Project." Washington, DC. http://documents.worldbank.org/curated/pt/753901468184440003/West-Bank-and-Gaza-Seventh-Palestinian-National-Development-Plan-Policy-Grant-Program-Project.

Woude, Joanne van der. 2008. "Why Maps Matter: New Geographies of Early American Culture." *American Quarterly* 60 (4): 1073–156.

Yiftachel, Oren. 1991. "State Policies, Land Control, and an Ethnic Minority: The Arabs in the Galilee Region, Israel." *Environment and Planning D: Society and Space* 9 (3): 329–62. https://doi.org/10.1068/d090329.

Yiftachel, Oren. 1996. "The Internal Frontier: Territorial Control and Ethnic Relations in Israel." *Regional Studies* 30 (5): 493–508.

Yiftachel, Oren. 1997. "Israeli Society and Jewish-Palestinian Reconciliation: 'Ethnocracy' and Its Territorial Contradictions." *Middle East Journal* 51 (4): 505–19.

Yiftachel, Oren. 2006. *Ethnocracy: Land and Identity Politics in Israel/Palestine*. Philadelphia: University of Pennsylvania Press.

Yiftachel, Oren. 2009a. "Theoretical Notes On 'Gray Cities': The Coming of Urban Apartheid?" *Planning Theory* 8 (1): 88–100.

Yiftachel, Oren. 2009b. "Critical Theory and 'Gray Space': Mobilization of the Colonized." *City* 13 (2–3): 246–63.

Yiftachel, Oren, and Alexandre (Sandy) Kedar. 2000. "Landed Power: The Making of the Israeli Land Regime." *Theory and Criticism* 16 (Spring): 67–100.

Young, James Edward. 1993. *The Texture of Memory: Holocaust Memorials and Meaning*. New Haven, CT: Yale University Press.

Yuchtman-Yaar, Ephraim, and Ze'ev Shavit. 2004. "The Cleavage between Jewish and Arab Israeli Citizens." In *Jews in Israel: Contemporary Social and Cultural Patterns*, edited by Uzi Rebhun and Chaim Isaac Waxman, 345–70. London: University Press of New England.

Zelinsky, Wilbur. 1984. "O Say, Can You See? Nationalistic Emblems in the Landscape." *Winterthur Portfolio* 19 (4): 277–86.

Zelinsky, Wilbur. 1986. "The Changing Face of Nationalism in the American Landscape." *Canadian Geographer* 30 (2): 171–75.

Zink, Valerie. 2009. "A Quiet Transfer: The Judaization of Jerusalem." *Contemporary Arab Affairs* 2 (1): 122–33.

Index

Note: Figures are indicated by *f* following the page number